Fiber Optic Cabling

Fiber Optic Cabling

Second Edition

Barry Elliott
Mike Gilmore

Newnes
Oxford Auckland Boston Johannesburg Melbourne New Delhi

Newnes
An imprint of Butterworth-Heinemann
Linacre House, Jordan Hill, Oxford OX2 8DP
225 Wildwood Avenue, Woburn, MA 01801-2041
A division of Reed Educational and Professional Publishing Ltd

℞ A member of the Reed Elsevier plc group

First published 1991
Second edition 2002

© Mike Gilmore and Barry Elliott 2002

All rights reserved. No part of this publication may be reproduced in any material form (including photocopying or storing in any medium by electronic means and whether or not transiently or incidentally to some other use of this publication) without the written permission of the copyright holder except in accordance with the provisions of the Copyright, Designs and Patents Act 1988 or under the terms of a licence issued by the Copyright Licensing Agency Ltd, 90 Tottenham Court Road, London, England W1P 0LP. Applications for the copyright holder's written permission to reproduce any part of this publication should be addressed to the publishers

British Library Cataloguing in Publication Data
A catalogue record for this book is available from the British Library

ISBN 0 7506 5013 3

Composition by Scribe Design, Gillingham, Kent, UK
Printed and bound in Great Britain by Biddles Ltd, *www.biddles.co.uk*

FOR EVERY VOLUME THAT WE PUBLISH, BUTTERWORTH-HEINEMANN
WILL PAY FOR BTCV TO PLANT AND CARE FOR A TREE.

Contents

	Preface	xi
	Abbreviations	xiii
1	**Fiber optic communications and the data cabling revolution**	**1**
	Safety statement	1
	Cabling as an operating system	1
	Communications cabling and its role	2
	Fiber optics and the cabling market	3
	Fiber optic cabling as an operating system	7
	The economics of fiber optic cabling	9
2	**Optical fiber theory**	**11**
	Introduction	11
	Basic fiber parameters	11
	Refractive index	12
	Laws of reflection and refraction	15
	Optical fiber and total internal reflection	18
	Optical fiber construction and definitions	20
	The ideal fiber	21
	Light acceptance and numerical aperture	22
	Light loss and attenuation	24
	Intrinsic loss mechanisms	24
	Modal distribution and fiber attenuation	27
	Extrinsic loss mechanisms	28
	Impact of numerical aperture on attenuation	31
	Operational wavelength windows	31
	Bandwidth	31
	Step index and graded index fibers	34
	Modal conversion and its effect upon bandwidth	36
	Single mode transmission in optical fiber	39

	Bandwidth specifications for optical fiber	45
	System design, bandwidth utilization and fiber geometries	46
	Optical fiber geometries	47
	The new family of single mode fiber	48
	Plastic optical fiber	52
	References	53
3	**Optical fiber production techniques**	54
	Introduction	54
	Manufacturing techniques	54
	Preform manufacture	55
	Stepped index fiber preforms	55
	All-silica fiber preforms	56
	Fiber manufacture from preforms	63
	Fiber compatibility	66
	Clad silica fibers	66
	Plastic optical fiber	67
	Radiation hardness	68
	Primary coating processes	70
	Summary	71
4	**Optical fiber – connection theory and basic techniques**	72
	Introduction	72
	Connection techniques	72
	Connection categories	73
	Insertion loss	73
	Basic parametric mismatch	74
	Fusion splice joints	78
	Mechanical alignment	79
	Total loss, fiber geometry and preparation	84
	Return loss	84
	Summary	87
5	**Practical aspects of connection technology**	88
	Introduction	88
	Alignment techniques within joints	88
	The joint and its specification	90
	Insertion loss and component specifications	91
	The introduction of optical fiber within joint mechanisms	95
	Joint mechanisms: relative cladding diameter alignment	98
	Joint mechanisms: absolute cladding diameter alignment	100
6	**Connectors and joints, alternatives and applications**	104
	Introduction	104

	Splice joints	105
	Demountable connectors	110
	Standards and optical connectors	121
	Termination: the attachment of a fiber optic connector to a cable	124
	Termination as an installation technique	127
	Summary	130
	References	131
7	**Fiber optic cables**	132
	Introduction	132
	Basic cabling elements	132
	Cabling requirements and designs	134
	Fiber optic cable design definitions	135
	Inter-building (external) cables	138
	Intra-building (internal) cables	141
	Fiber optic cables and optomechanical stresses	143
	User-friendly cable designs	147
	The economics of optical fiber cable design	147
	Summary	151
8	**Optical fiber highways**	152
	Introduction	152
	Optical fiber installations: definitions	152
	The optical fiber highway	154
	Optical fiber highway design	156
9	**Optical fiber highway design**	158
	Introduction	158
	Highway topology	158
	Nodal design	168
	Service needs	172
	Optical budget	176
	Bandwidth requirements	185
	Fiber geometry choices within the highway design	189
	Summary	192
	Reference	192
10	**Component choice**	193
	Introduction	193
	Fiber optic cable and cable assemblies	193
	Connectors	199
	Splice components	200
	Termination enclosures	201

Summary 204

11 Specification definition 205
Introduction 205
Technical ground rules 205
Operational requirement 206
Design proposal 211
Optical specification 214
Contractual aspects of the specification agreement 215
Summary 218

12 Acceptance test methods 219
Introduction 219
Fixed cables 219
Air-blown fiber testing 229
Cable assembly acceptance testing 229
Direct termination during installation and its effect upon quality assurance 239
Termination enclosures 239
Pre-installed cabling 240
Short-range systems and test philosophies 240

13 Installation practice 242
Introduction 242
Transmission equipment and the overall contract requirement 243
The role of the installer 244
The typical installation 244
Contract management 245
Installation programme 248
Termination practices 253

14 Final acceptance testing 256
Introduction 256
General inspection 256
Optical performance testing 259
Overall span attenuation measurement 262
Optical time domain reflectometer testing of installed spans 267

15 Documentation 274
Introduction 274
Contract documentation 274
Technical documentation 275
The function of final highway documentation 283
International standards concerning project documentation 283

16	**Repair and maintenance**	286
	Introduction	286
	Repair	286
	Maintenance	289
	Summary	290
17	**Case study**	291
	Introduction	291
	Network requirements	291
	Preliminary ideas	291
	Initial implementation for inter-building cabling	292
	Materials choice	300
	Bill of materials (fiber optic content)	304
	Installation planning	309
	Summary	309
18	**Future developments**	310
	Introduction	310
	Exotic lasers	310
	New optical fibers	311
	Next generation components	312
	New coding techniques	313

Appendix A   Attenuation within optical fiber: its measurement   314

Index   317

Preface

Mike Gilmore wrote the first edition of this book, the first major work on practical data communications optical fibers, in 1991. Mike has since become one of the most respected consultants in the field of structured/premises cabling in Europe and is the UK national expert: it thus falls on me to have the honour of being able to update this book in 2001, after ten years of unparalleled and dramatic growth in the optical communications industry.

In 2000, world production of optical fiber grew to 105 million kilometres, itself a 300% growth over the second half of the last decade. Optical fiber has become the undisputed medium of choice for long-haul telecommunications systems and is even delivered direct to many larger businesses. Trials are under way in Scandinavia and America to put fiber into the home to judge the true economics of the competing broadband technologies that will inevitably be delivered to every household.

The choice between different kinds of single mode fiber and the network topology it sits within are business critical decisions for the telecommunications network provider. The deregulation of the telecommunications markets in most countries has led to an explosion of growth in new carriers and an insatiable demand for optical fiber and components such as wavelength division multiplexers.

This book, however, focuses upon the use of optical fiber in data communications, local area networks and premises cabling. This is an area traditionally seen as 'lower-tech' where lower-performance multimode fiber was the order of the day. This was mostly true up until about 1997. Before that, multimode fiber with an SC or ST connector on the end would happily transport 100 Mb/s of data across a 2 kilometre campus. Beyond 2 kilometres was the world of telecommunications. The advent of gigabit Ethernet brought the 'event horizon' of single mode fiber down to the 500 metre mark. The arrival of ten gigabit Ethernet brings single mode all the way down to below 300 metres. At ten gigabit speeds the worlds of data communications and telecommunications are merging. With a new generation of Small Form Factor optical connectors to consider as well as an unknown mix of multimode and single mode fibers, campus optical cabling has suddenly got interesting again and nearly approaches the pioneering spirit of 1991 where the use of optical fiber on a campus was often seen as an act of faith, certainly in the choice of installer anyway.

One major change since 1991 has been the arrival of international standards that define nearly every detail of component performance, network design and system testing. The standards work is led by ANSI/TIA/EIA in America, by CENELEC in Europe and ITU and ISO/IEC for the rest of the world. All the appropriate standards are referred to in this edition along with the performance, selection and testing of all cables and components likely to be encountered in the LAN cabling environment.

Fiber-to-the-desk has not met the promises of the early 1990s. Some people say that copper cable has got better, with twisted-pair Category 5 and 6 copper cables offering frequency ranges up to 250 MHz. Copper cable hasn't changed that much; Shannon demonstrated mathematically the information carrying capacity of communications channels, including copper cables, in the 1930s. What has changed is the arrival of cheap digital signal processing power that enables exotic coding schemes to fully exploit the inherent bandwidth of well-made copper cables. Such microprocessors would simply not have been available or affordable in the early 1990s.

Today, fiber-to-the-desk is the preserve of those organizations that really need the extra benefits of optical fiber, such as longer transmission runs (copper horizontal cabling is limited to 100 metres) and those who want the security of optical fiber transmission, hence the popularity of fiber-to-the-desk solutions within the military. Fiber tends to get cheaper, as do the latest connectors and especially the optical transmission equipment, which for too long has been a major barrier to the uptake of short-distance optical fiber runs. Copper cable tends to get more expensive as the electrical demands upon it get higher and higher, while other factors such as the need to remotely power IP telephones over the cabling add yet more ingredients to an already complex technical/economic argument.

In Mike Gilmore's original book the last chapter was devoted to 'future developments'. All of his predictions have mostly come to pass and I finish this edition with my predictions of the future. For a book written in 2001 it is perhaps appropriate to quote *the* great technical prophet, Arthur C. Clarke, who wrote in 1975:

> *The only uncertainty, and a pretty harrowing one to the people who have to make decisions, is how quickly coaxial cables are going to be replaced by glass fibers, with their millionfold greater communications capability.*

Barry Elliott
2001: Credo ut intelligam

Abbreviations

ABF	Air Blown Fiber
ANSI	American National Standards Institute
APC	Angled Physical Contact
ATM	Asynchronous Transfer Mode
BER	Bit Error Rate
CATV	Community Antenna Television (cable TV)
CCI	Core Cladding Interface
COA	Centralized Optical Architecture
CPD	Construction Products Directive
CWDM	Coarse Wavelength Division Multiplexing
DFB	Distributed Feedback (laser)
DMD	Differential Modal Delay
DSF	Dispersion Shifted Fiber
DWDM	Dense Wavelength Division Multiplexing
dB	decibel
EDFA	Erbium Doped Fiber Amplifier
EF	Encircled Flux
EIA	Electronic Industries Alliance
EMB	Effective Modal Bandwidth
EMC	Electro Magnetic Compatibility
EMI	Electro Magnetic Immunity (or sometimes 'EM *Interference*')
ESD	Electro Static Discharge
FCC	Federal Communications Commission
FDDI	Fiber Distributed Data Interface
FP	Fabry Perot (laser)
FDM	Frequency Division Multiplexing
FOCIS	Fiber Optic Connector Intermateability Standard
GHz	Gigahertz
GI	Graded Index

GPa	Giga Pascal
HCS	Hard Clad Silica
HPPI	High Performance Parallel Interface
ICEA	Insulated Cable Engineers Association
IEC	International Electro Technical Commission
IEE	Institute of Electrical Engineers (UK)
IEEE	Institute of Electrical and Electronic Engineers (USA)
ISDN	Integrated Services Digital Network
ISO	International Standards Organization
ITU	International Telecommunications Union
IVD	Inside Vapour Deposition
LAN	Local Area Network
LEAF	Large Effective Area Fiber
LED	Light Emitting Diode
LFH	Low Fire Hazard
LSZH	Low Smoke Zero Halogen
MAN	Metropolitan Area Network
Mb/s	Megabits per second
MCVD	Modified Chemical Vapour Deposition
MEMS	Micro Eectro Mechanical Systems
MHz	Megahertz
NA	Numerical Aperture
nm	Nanometres
NEC	National Electrical Code (USA)
NEMA	National Electrical Manufacturers Association (USA)
NRZ	Non-Return to Zero
NTT	Nippon Telephone and Telegraph
NZDS	Non Zero Dispersion Shifted (fiber)
OFL	Overfilled Launch
OVD	Outside Vapour Deposition
PAM	Pulse Amplitude Modulation
PC	Physical Contact
PCOF	Primary Coated Optical Fiber
PCS	Plastic Clad Silica
PCVD	Plasma Chemical Vapour Deposition
PMD	Polarization Mode Dispersion
PMMA	Poly Methyl Methcrylate
POF	Plastic Optical Fiber
PTFE	Poly Tetra Fluoro Ethylene
PTT	Public Telephone and Telegraph (operator)
PVC	Poly Vinyl Chloride
OCDMA	Optical Code Division Multiple Access
RML	Restricted Mode Launch
SAN	Storage Area Network

SC	Subscriber Connector	
SCOF	Secondary Coated Optical Fiber	
SCSI	Small Computer System Interface	
SFF	Small Form Factor (optical connectors)	
SMA	Sub Miniature Assembly	
SMF	Single Mode Fiber	
SNR	Signal to Noise Ratio	
SoHo	Small Office Home Office	
SONET	Synchronous Optical Network	
SROFC	Single Ruggedized Optical Fiber Cable	
TDM	Time Division Multiplexing	
TIA	Telecommunications Industry Association	
TIR	Total Internal Reflection	
TO	Telecommunications Outlet	
TSB	Telecommunications Systems Bulletin	
UL	Underwriters Laboratory	
VAD	Vapour Axial Deposition	
VCSEL	Vertical Cavity Surface Emitting Laser	
WAN	Wide Area Network	
WDM	Wavelength Division Multiplexing	
WWDM	Wide Wavelength Division Multiplexing	

1 Fiber optic communications and the data cabling revolution

Safety statement

If you are reading this book then it means you have a practical interest in the use of optical fiber. You should be aware of the safety issues concerning the handling of optical fiber and its accessories.

- Always dispose of optical fiber off-cuts in a suitable 'sharps' container.
- Never look into the end of fiber optic equipment, devices or fibers unless you know what they are connected to. They may be emitting invisible infrared radiation which may be injurious to the eyes.
- Optical connector terminating ovens are hot and may give off fumes that are irritants to some people.

Cabling as an operating system

Information technology is an often used, and misused, term. It encompasses a bewildering array of concepts and there is a tendency to pigeonhole any new electronics or communications technology or product as a part of the information technology revolution.

Certainly from the viewpoint that most electronic hardware incorporates some element of communication with itself, its close family or with ourselves, then it is possible to include virtually all modern equipment under the high-technology, information-technology banner. What is undeniable is that communications between persons and between equipment is facing an incredible rate of growth. Indeed new forms of communication arrive on the market so regularly that for most people any detailed understanding is impossible. It may be positively undesirable to investigate too deeply since it is likely that subsequent generations of equipment would render any previously gained expertise rather redundant. It is tempting therefore to dismiss the entire progression as the impact of information technology. Never has it been more enticing to become a jack-of-all-trades believing that the master of one is destined to fail. Under these circumstances the most important factor is the ability of the user to be able to use, rather than understand, the various systems. At the most basic level this means that it is more desirable to be able to use a telephone than it is to be familiar with the intricacies of exchange-switching components.

As computers have evolved the standardization of software-based operating systems has assisted their acceptance in the market because the user feels more relaxed and less intimidated by existing and new equipment. This concentration upon operation rather than technical appreciation is reflected in the area of communications cabling. Until recently the cabling between various devices within a communications network (e.g. computer and many peripherals) was an invisible product, and cost, to the customer. Indeed many customers were unaware of the routing, capability and reliability of the cabling which, to a great extent, was responsible for the continuing operation of their network.

More recently, however, a gradual revolution has taken place and the cabling network linking the various components within the communications system has become the hardware equivalent of the software operating system. Rather than being specific to the two pieces of equipment at either end of the cable the installed cabling supports the use of many other devices and peripherals. As such the cabling is an operational issue rather than a technical one and involves general management decisions in addition to those made on engineering grounds.

The cabling philosophy of a company is now a central communications issue and represents a substantial investment not merely supporting today's equipment (and its processing requirements) but to service a wide range of equipment for an extended period of time. As such the cabling is no longer an invisible overhead within a computer-package purchase but rather a major capital expense which must show effective return on investment and exhibit true extended operational lifetime.

Communications cabling and its role

Communication between two or more communicators can be achieved in a variety of ways but can always be broadly categorized as follows:

- the type of communicated data: e.g. telephony, data communication, video transmission;
- the importance of the communicated data;
- the environment surrounding the communicated data: e.g. distance, bandwidth, electromagnetic factors including security, electrical noise etc.

Historically the value of the communicated data was much less crucial than it is now or will be in the future. If a domestic or office telephone line failed then voice data was interrupted and alternative arrangements could be made. However, if a main telecommunications link fails the cost can be significant both in terms of the data lost at the moment of failure and, more importantly, the cost of extended downtime. When analysed it is easy to see that this trend towards ever more important communicated data has resulted from

- the rapid spread in the use of computing equipment;
- the increased capacity of the equipment to analyse and respond to communicated information.

These two factors have resulted in physically extended communication networks operating at higher speeds. In turn this has led to an increased use of interconnecting cable. The impact of the failure of these interconnections depends upon the value of the data interrupted.

The concept of an extended cabling infrastructure is therefore no longer a series of 'strands of wire' linking one component with another but is rather a carefully designed network of cables (each meeting its own technical specification) installed to provide high-speed communication paths which have been designed to be reliable with minimal mean-time-to-repair figures.

Communications cabling has become a combination of product specification (cable) and network design (repair philosophy, installation practice) consistent with its importance. This concept separates the cabling from the transmission hardware and suggests a close analogy with the concept of the computer operating system and its independence from user generated software packages. This book concentrates upon the use of optical fiber as a transmission medium within the cabling system and as indicated above does not require knowledge of individual communication protocol or transmission equipment.

Fiber optics and the cabling market

Telecommunications

The largest communications network in any country is the public telecommunications network. Cabling represents the vast majority of the total investment applied to these frequently complex transmission paths. Accordingly the relevant authorities and highly competitive, newly deregulated telcos are always at the forefront of technological changes, ensuring that growth in communication requirements (generated by either population increase or the 'information technology revolution') can be met with least additional cost of ownership.

A telecommunications network may therefore be considered to be the foremost cabling infrastructure and the impact of new technology can be expected to be examined first in this area of communications.

In 1966 Charles Kao and George Hockham (Standard Telephone Laboratories, Harlow, England) announced the possibility of data communication by the passage of light (infrared) along an optically transmissive medium. The telecommunications authorities rapidly reviewed the opportunity and the potential advantages were found to be highly attractive.

The transmission of signal data by passing light signals down suitable optical media was of interest for two main reasons (considered to be the two primary advantages of optical fiber technology): high bandwidth (or data-carrying capacity) and low attenuation (or power loss).

Bandwidth is a measure of the capacity of the medium to transmit data. The higher the bandwidth, the faster the data can be injected whilst maintaining acceptable error rates at the point of reception. For the telecommunications industry the importance was clear; the higher the bandwidth of the transmission medium, the fewer individual transmitting elements that are needed. Optical fiber elements boast tremendously high bandwidths and their use has drastically reduced the size of cables whilst increasing the data-carrying capacity over their bulkier copper counterparts. This factor is reinforced by a third advantage: optical fiber manufactured from either glass or, more commonly, silica is an electrically non-conductive material and as such is unaffected by crosstalk between elements. This feature removes the need for screening of individual transmission elements, thereby further reducing the cable diameters.

With particular regard to the telecommunications industry it was also realized that if fewer cabled elements were required then fewer individual transceivers would be needed at the repeater/regenerator stations. This not only reduces costs of installation and ownership of the network but also increases reliability. The issue of repeater/regenerators was particularly relevant since the second primary advantage of optical fiber is its very low signal–power attenuation. This obviously was of interest to the telecommunications organizations since it suggested the opportunity for greater inter-repeater distances. This suggested lower numbers of repeaters, again leading to lower costs and increased reliability.

The twin ambitions of lower costs and increased reliability were undoubtedly attractive to the telecommunications authorities but the main benefit of optical fiber, in an age of rapid growth in communications traffic, was, and still is, bandwidth. The fiber optic cables now installed as trunk and local carriers within the telecommunications system are not a limiting factor in the level of services offered. It is actually more correct to say that capacity is limited by the capability of light injection and detection devices.

It is worthwhile to point out that the reductions in cost indicated above did not occur overnight and multi-million pound investments were undertaken by the fiber optics industries to develop the product to its current level of performance. However, the costing structure that existed by 2001 is an excellent example of high-technology product development linked to volume production with resultant large-scale cost reductions. The large volume of component usage in the telecommunications industry is directly responsible for this situation and the rapid growth of alternative applications is based upon the foundations laid by the industry.

As a result it is now possible to purchase, at low cost, the high specification components, equipment and installation technology to service the growing volume market in the data communications sector discussed in detail below.

Military communications

At the time optical fiber was first proposed as a means of communication the advantages to telecommunications were immediately apparent. The fundamental advantages of high bandwidth, low signal attenuation and the non-conducting nature of the medium placed optical fiber in the forefront of new technology within the communications sector.

However, much early work was also undertaken on behalf of the defence industry. A large amount of development effort was funded with the aim of designing and manufacturing a variety of components suitable for further integration into the fiber optic communication systems specific to the military arena. Applications in land-based field communications systems and shipborne and airborne command and control systems have generated a range of equipment which is totally different in character from that needed in telecommunications systems. The benefits of bandwidth and signal attenuation, dominant in the telecommunications area, were less important in the military markets. The secondary benefits of optical fiber such as resistance to electromagnetic interference, security and cable weight (and volume) were much more relevant for the relatively short-haul systems encountered. The result of this continuing involvement by the military sector has been the creation of a range of products capable of meeting a wide range of cabling requirements – primarily at the opposite end of the technical spectrum from telecommunications but no less valid.

Unfortunately much of the early work did not result in the full-scale production of fiber optic systems despite the basic work being broadly successful. The fundamental reason for this is that in many cases the fiber optic system was considered to be merely an alternative to an existing copper cabling network, justifiable only on the grounds of secondary issues such as security, weight savings etc. In no way were these systems utilizing the main features of optical fiber technology, bandwidth and attenuation, which could not be readily attained by copper. The high price of the optical variant frequently led to the subtle benefits offered by fiber being adjudged to be not cost effective.

More recently the future-proof aspects of optical fiber technology have been seen to be applicable to military communications. Since the communications requirements within all the fighting services have been observed to be increasing broadly in line with those in the commercial market it has become necessary to provide cabling systems which exceed the capacity of copper technology.

In many cases therefore the technology now adopted owes more to the components of telecommunications rather than the early military developments but in formats and structures suitable for the military environment.

Although fiber-to-the-desk has been heralded as 'next year's technology' in the data communications industry, it is the military sector which has become the most enthusiastic proponent of fiber-to-the-desk solutions, precisely for reasons of security.

The data communications market

The term 'data communications' is generally accepted to indicate the transfer of computer-based information as opposed to telecommunications which is regarded as being the transfer of telephonic information. This is indeed a fine distinction and in recent years the separation between the two types of communications has become ever more blurred as the two technologies have been seen to converge.

Nevertheless the general opinion is that data communications is the transfer of information which lies outside the telecommunications networks and as such is generally regarded as being linked to the local area network (LAN) and building cabling markets. This broad definition is accepted within this book. The term 'local area network' is also rather vague but includes many applications within the computer industry, military command and control systems together with the commercial process-control markets.

Having briefly discussed in the preceding section the evolution of fiber for data communications within the military sector, it is relevant to separately review its application to commercial data communications.

As discussed above, the long-term cost effectiveness of optical fiber was of interest to the telecommunications industry because the cabling infrastructure was treated as a major asset having a significant influence over the reliability of the entire communications system. For the more localized topologies of commercial data networks the actual cabling received little interest or respect for three main reasons:

- The amount of data transmitted was generally much lower.
- Usage of data was more centralized.
- Growth in transmission requirements was generally more restricted.

It is hardly surprising therefore that a new medium offering wideband transmission over considerable distances tended to meet commercial resistance due to its cost. However, a number of prototype or evaluation systems were installed in the latter half of the 1970s which were matched by a significant amount of development work in the laboratories of the major communications and computing organizations. The more advanced of these groups produced fiber optic variants of their previously all-copper systems in preparation for the forecast upturn in data communications caused by the information-technology revolution.

As a result of this revolution the amount of data transmitted has increased to an undreamed degree and, perhaps more importantly, is expected to continue to increase at an almost exponential rate as computer peripherals become ever more complex, thereby offering new services needing faster communication. The three decades between 1970 and 2000 have demonstrated a growth in LAN speed of about a factor of 100-fold per decade. Also the distribution of the information has grown as developments have allowed the sharing of computing power across large manufacturing sites or within office complexes.

These changes together with the reduction in cost of fiber optic components generated by the telecommunications market have now led to a rapidly increasing use of the technology within the 'data communications' market. Consequently the data communications market had historically chosen optical fiber on a limited basis. More recently transmission requirements have finally grown to a level which favours the application of optical fiber for similar reasons to those seen in telecommunications, with its use justified by virtue of its bandwidth, servicing both immediate and future communications requirements.

The growth in standardized structured cabling systems has seen optical fiber firmly established as the preferred medium for building backbone and campus cabling applications; indeed it is now the only media that could transport multi-gigabit traffic.

Fiber optic cabling as an operating system

The above section briefly discussed the history of the uptake of optical fiber as a cabling medium in telecommunications, military and data communications.

It is clear, however, that as the information transfer requirements have grown in the non-telecommunications sector, so the solutions for cabling have become more linked to those adopted for telecommunications. This is quite simply because organizations are viewing even small communications networks as comprising transmission equipment and, but separate from, the cabling medium itself.

The cabling medium, be it copper or optical, is now frequently seen as a separate capital investment which will only be truly effective if it can be seen to support multiple upgrades in transmission hardware without any need to reinstall the cabling.

The advent of communications standards such as the IEEE 802.x systems (Ethernet, token ring etc.) has led to the standardization of cabling to support the various protocols. This approach to 'communication-standards' cabling justifies the concept of cabling as an operating system. The 10 megabits per second (10 Mb/s) copper Ethernet and IBM token ring (4 Mb/s and 16 Mb/s) cabling can support transmission requirements well beyond those which were considered typical during the early 1980s. However, even as copper cable transmission speeds ramp up to 1000 Mb/s (gigabit Ethernet over Cat 5e) and potentially 2.5 gigabit Ethernet and 2.4 Gb/s ATM over Cat 6, copper cable is still going to be limited by distance and EMC problems. This is coupled to the fact that as copper cabling becomes more complex, it becomes more expensive, whereas fiber cabling and components get relatively cheaper every year. Optical fiber is the medium to be adopted which offers extended operational lifetime. People should always be wary of terms such as 'future-proof', however. The 1980s and 1990s were typified by LAN installations consisting of medium quality 62.5/125 multimode fiber being installed in the backbone, on the selling slogan, 'it's optical fiber, it must be future-proof'. The advent of gigabit and ten gigabit Ethernet has shown that 62.5/125 fiber has long since run out of steam in backbone applications, and what were once 2000 metre backbones supporting 100 Mb/s, have now been reduced to fifty metres or less when trying to cope with ten gigabit Ethernet. Only single mode fiber, with its near infinite bandwidth, can ever be described as future-proof.

In many applications an optical fiber solution represents the ultimate operating system offering the user operational lifetimes in excess of all normal capital investment return profiles (five, seven or even ten years).

The majority of capital-based cabling networks are now designed, having considered the application of optical fiber as either part or all of their cabling operating system. In doing this, the designers are effectively adopting the telecommunication solution to their cabling requirements. Interestingly the specific optical components (and their technological generation) adopted within the short-haul data-communications market are generally those originally used within the trunk telecommunications networks of the early 1980s, whereas the future of all fiber communications is based upon the telecommunications market as it moves into the short-haul, local-loop subscriber connection. In this way the convergence between computing and telecommunications is heavily underlined.

The economics of fiber optic cabling

Since its first proposal in 1966 the economics behind optical fiber technology have changed radically. The major components within the communications system comprise the fiber (and the resulting cable), the connections and the opto-electronic conversion equipment necessary to convert the electrical signal to light and vice versa.

In the early years of optical transmission the relatively high cost of the above items had to be balanced by the savings achieved within the remainder of the system. In the case of telecommunications these other savings were generated by the removal of repeater/regenerator stations. Thus the concept of 'break-even' distance grew rapidly and was broadly defined as the distance at which the total cost of a copper system would be equivalent to that of the optical fiber alternative. For systems in excess of that length the optical option would offer overall cost savings whereas shorter-haul systems would favour copper – unless other technical factors overrode that choice.

It is not surprising therefore that long-range telecommunications was the first user group to seriously consider the optical medium. Similarly the technology was an obvious candidate in the area of long-range video transmission (motorway surveillance, cable and satellite TV distribution). The cost advantages were immediately apparent and practical applications were soon forthcoming.

Based upon the volume production of cable and connectors for the telecommunications market the inevitable cost reductions tended to reduce the 'break-even' distance.

When the argument is purely on cost grounds it is a relatively straightforward decision. Unfortunately even when the cost of cabling is fairly matched between copper and fiber optics the additional cost of opto-electronic converters cannot be ignored. Until certain key criteria are met the complete domination of data communications by optical fiber cannot be achieved or even expected.

These criteria are as follows:

- standardization of fiber type such that telecommunications product can be used in all application areas;
- reductions in the cost of opto-electronic converters based upon large volume usage;
- a widespread requirement for the data transmission at speeds which increase the cost of the copper medium or, in the extreme, preclude the use of copper totally.

These three milestones are rapidly being approached; the first two by the application of fiber to the telecommunications subscriber loop (to the home) whilst the third is more frequently encountered due to vastly increased needs for services.

Meanwhile the economics of fiber optic cabling dictate that while 'break-even' distances have decreased the widespread use of 'fiber-to-the-desk' is still some time away.

There is a popular misconception in the press that the 'fiber optic revolution' has not yet occurred. It is evidently assumed that the revolution is an overnight occurrence that miraculously converts every copper cabling installation to optical fiber. This is rather unfortunate propaganda and, to a great extent, both untrue and unrealistic.

In telecommunications, optical fiber carries information not only in the trunk network but also to the local exchanges. For motorway surveillance the use of optical fiber is mandatory in many areas. At the data communications level all the major computer suppliers have some fiber optic product offering within their cabling systems. Increasingly process control systems suppliers are able to offer optical solutions within large projects.

But in most, if not all, cases the fiber optic medium is not a total solution but rather a partial, more targeted, solution within an overall cabling philosophy. There is no 'fiber optic revolution' as such. There is instead a carefully assessed strategy offering the user the services required over the media best suited to the environment.

What cannot be ignored is the fact that fiber optic cabling is specifically viewed as a future-proofed element in the larger cabling market and as such operates more readily as an operating system deserving deep consideration at the design, installation, documentation and post-installation stages.

As has been seen, the immediate cost benefits of adopting a total fiber optic cabling strategy are dependent upon the transmission distance. With the exception of telecommunications and long-haul surveillance systems the typical dimensions of communications networks are quite limited.

The local area network is frequently defined as having a 2 kilometre span. The vast majority of fiber optic cabling within the data communications market will have links that do not exceed 500 metres. Such networks, when installed using professional grades of optical fiber, offer enormous potential for upgrades in transmission equipment and services. The choice of components, network topologies, cabling design, instal_lation techniques and documentation are all critical to the establishment of a cabling network which maximizes the operational return on investment.

The remainder of this book deals with these topics individually whilst building in a modular fashion to ensure that fiber optic cabling networks most fully meet their potential as operating systems.

2 Optical fiber theory

Introduction

The theory of transmission of light through optical fiber can undoubtedly be treated at a number of intellectual levels ranging from the highly simplistic to the mathematically complex. During the frequent specialist training courses operated by the authors the delegates are advised that 11th grade (GCSE) level physics and basic trigonometry are the only tools required for a comprehensive understanding of optical fiber, its parameters and its history. That being said it does help if one can grasp the concept of light as being a ray, a particle and a wave – though thankfully not all at the same time.

This chapter reviews the theory of transmission of light along an optical medium from the viewpoint of cabling design and practice rather than theoretical exactitude.

As perhaps the most important chapter of the book, it is intended to give the reader a working knowledge of transmission theory as it relates to products currently available. It forms a basis for the understanding of loss mechanisms throughout installed networks and, perhaps more importantly, it allows the reader to establish the validity of a proposed fiber optic cabling installation as an operating system based upon its bandwidth (or data capacity).

Basic fiber parameters

Optical fiber transmission is very straightforward. There are only two reasons why a particular system might not operate:

- poor design of, or damage to, the transmission equipment;
- poor design of, or damage to, the interconnecting fiber and components.

Equally simply there are just three basic reasons why a particular interconnection might not operate:

- insufficient light launched into the fiber;
- excessive light lost within the fiber;
- insufficient bandwidth within the fiber.

At the design stage the basic parameters of an optical fiber can be considered to be:

- light acceptance;
- light loss;
- bandwidth.

It will be seen that all three parameters are governed by two other more basic factors: these are the active diameter of the fiber and the refractive indices of the materials used within the fiber. The analysis of the operation of optical fiber can thus be reduced to the understanding of a very few basic concepts.

Refractive index

All materials that allow the transmission of electromagnetic radiation have an associated refractive index. In copper cables this is analogous to the NVP or nominal velocity of propagation.

This refractive index is denoted by n and is defined by the equation (2.1):

$$n = \frac{\text{velocity of light in a vacuum}}{\text{velocity of light in the medium}} \tag{2.1}$$

As light travels through a vacuum uninterrupted by any material structure it is logical to assume that the velocity of light in a vacuum is the highest achievable value. In all other materials the light is interrupted to a lesser or greater extent by the atomic structure of that material and as a result will travel more slowly.

Therefore the refractive index of a vacuum is unity (1.0) and all other media have refractive indices greater than unity. Table 2.1 provides some general information with regard to refractive index and velocities of light in various materials.

The refractive index of materials used within an optical fiber have a direct influence upon the basic properties of the fiber.

A more detailed analysis of refractive index mathematics shows that the index is not a constant value but instead depends upon the wavelength of light at which it is measured. Further it should be remembered that the term 'light' is not confined to the visible spectrum. The definition and measurement of refractive index is valid for all types of 'light', more fully defined as 'electromagnetic radiation'.

Table 2.1 *Typical refractive index values*

Material	Refractive index
Gases	
air	1.00027
Liquids	
water	1.333
alcohol	1.361
Solids	
pure silica	1.458
salt (NaCl)	1.500
amber	1.500
diamond	2.419

A more comprehensive definition of refractive index can be given as defined in equation (2.2):

$$n\lambda = \frac{\text{velocity of eletromagnetic radiation at wavelength } \lambda \text{ in a vacuum}}{\text{velocity of eletromagnetic radiation at wavelength } \lambda \text{ in the material}}$$

$$n\lambda = \frac{\text{constant for all } \lambda}{\text{variable with } \lambda} \qquad (2.2)$$

It can therefore be seen that the refractive index of a material may vary across the electromagnetic radiation spectrum. Figures 2.1 and 2.2 provide further information regarding the electromagnetic spectrum and Table 2.2

Table 2.2 *Pure silica: refractive index variation with wavelength*

Wavelength λ (nm)	Refractive index n
600	1.4580
700	1.4553
800	1.4533
900	1.4518
1000	1.4504
1100	1.4492
1200	1.4481
1300	1.4469
1400	1.4458
1500	1.4466
1600	1.4434
1700	1.4422
1800	1.4409

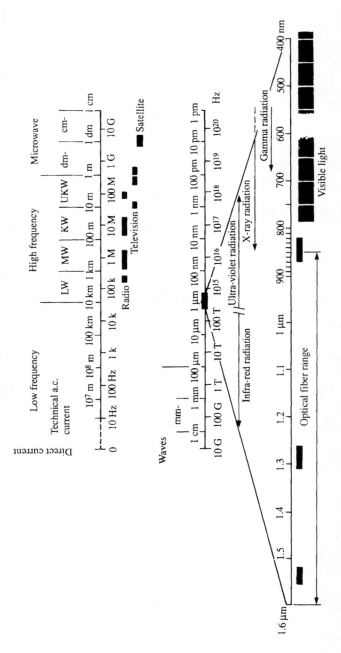

Figure 2.1 *Electromagnetic spectrum*

Optical fiber theory 15

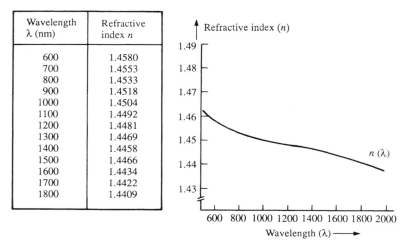

Figure 2.2 *Pure silica: refractive index variation with wavelength*

shows typical figures of refractive index against wavelength, together with the corresponding graph, for silica, the basic constituent of all professional-quality optical fibers.

Laws of reflection and refraction

Optical fiber transmission depends upon the passage of electromagnetic radiation, typically infrared light, along a silica or glass-based medium by the processes of reflection and refraction. To fully understand both the advantages and limitations of optical fiber it is necessary to review the simple laws of reflection and refraction of electromagnetic radiation.

Refraction

Refraction is the scientific term applied to the bending of light due to variations in refractive index. Refraction can be experienced in a large number of practical ways, including the following:

- the image of a pole immersed in a pond appears to bend at the surface of the water;
- 'mirages' appear to show distant images as being temptingly close at hand;
- spectacle or binocular lenses all manipulate light by bending in order to magnify or modify the images produced.

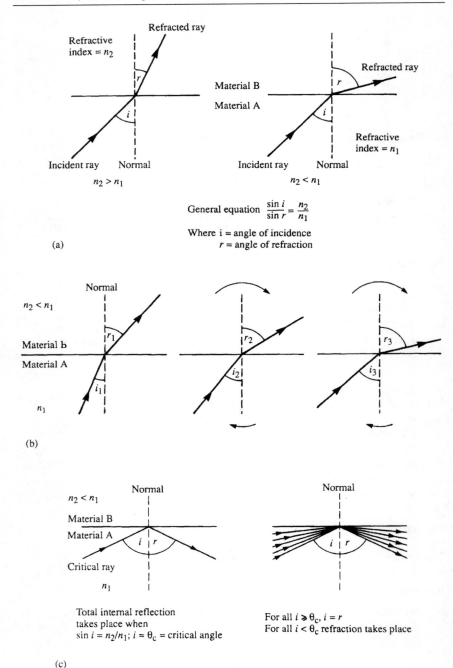

Figure 2.3 (a) Refraction of light; (b) rotation of incident and refracted rays; (c) total internal reflection

Figures 2.3 (a), (b) and (c) show the various stages of refraction as they apply to optical fiber

In Figure 2.3(a) the standard form of refraction is depicted. Two materials with different refractive indices are separated by a smooth interface AB. If a light ray X originates within the base material it will be refracted or bent at the interface. The direction in which the light is refracted is dependent upon the indices of the two materials. If n_1 is greater than n_2, then the ray X is refracted away from the normal whereas if n_1 is less than n_2, then the light is refracted towards the normal.

Refraction is governed by equation (2.3):

$$\frac{\sin i}{\sin r} = \frac{n_2}{n_1} \qquad (2.3)$$

When applying this equation to optical fiber then the case of n_1 greater than n_2 should be investigated. Light is refracted away from the normal. As the angle of incidence (*i*) increases so does the angle of refraction (*r*). Figure 2.3(b) shows this effect.

However, the angle of refraction cannot exceed 90°, for which sin *r* is unity. At this point the process of refraction undergoes an important change. Light is no longer refracted out of the base medium but instead it is reflected back into the base medium itself. The angle of incidence at which this effect takes place is known as the critical angle, denoted by θ_c, expressed in equation (2.4):

$$\sin \theta_c = \frac{n_2}{n_1} \qquad (2.4)$$

For all angles of incidence greater than the critical angle the light will be reflected back into the base medium due to this effect, which is called total internal reflection. The two key features of total internal reflection are that:

- The angle of incidence = the angle of reflection.
- There is no loss of radiated power at the reflection. This, put more simply, means that there is no loss of light at the interface and that, in theory at least, total internal reflection could take place indefinitely.

Figure 2.3 (c) shows the effect and the relevant equations.

Fresnel reflection

Before passing on to optical fiber and its basic theory it is useful to discuss a further type of reflection, Fresnel reflection. Fresnel reflection takes place where refraction is involved, i.e. where light travels across the interface between two materials having different refractive indices. Figure 2.4 demonstrates the effect and defines the equations for power levels resulting from

18 Fiber Optic Cabling

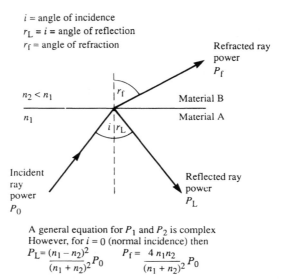

Figure 2.4 Fresnel reflections

the Fresnel reflection. It is clear from the equations in Figure 2.4 that the greater the difference in refractive index between the two materials then the greater is the strength of the reflection and therefore the associated power loss. It will also be noted that the loss occurs independently of the direction of the light path.

In general, light will be lost in the forward direction each time a refractive index barrier is traversed; however, it should be highlighted that when the angle of incidence is greater than θ_c, the critical angle, then total internal reflection takes place and there is no passage of light from one medium to the other and no reduction in forward transmitted power.

Optical fiber and total internal reflection

The phenomenon of total internal reflection (TIR) is not a new concept. Indeed all the equations detailed thus far in this chapter are forms of Snell's laws (of reflection and refraction) and were first outlined in 1621.

In the eighteenth century it was known that light could be guided by jets or streams of liquid since the high refractive index of the liquid contained the light as the streams passed through the air of low refractive index surrounding them. Nevertheless this observation appears a long way short of the complex technology required to transmit telecommunications information over many tens of kilometres of optical fiber.

Optical fiber theory 19

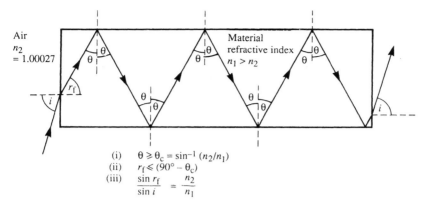

Figure 2.5 *Basic optical transmission*

This section discusses the manner in which total internal reflection is achieved in optical fiber and defines the various components involved. Figure 2.3(c) has already shown the basic characteristics of TIR. If a material of high refractive index were produced in a cylindrical format which would have and, more importantly, retain a smooth unblemished interface between itself and its surroundings of a lower refractive index (air = 1.00027) then it should be possible to create multiple TIR as shown in Figure 2.5.

This would in fact constitute a basic optical transmission element but unfortunately it has proved impossible to maintain the smooth, unblemished interface in air due to surface damage and contaminants. Figure 2.6 shows the impact of such surface irregularities.

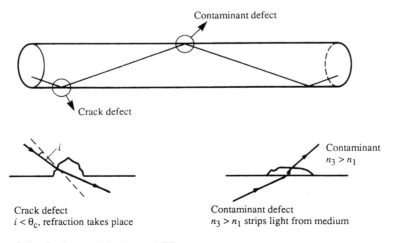

Figure 2.6 *Surface defects and TIR*

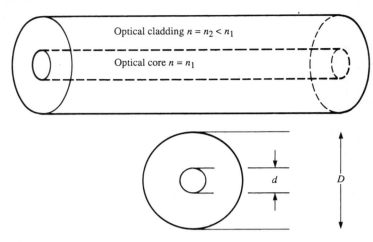

Figure 2.7 Core—cladding arrangement

It is therefore necessary to achieve and maintain the interface surface quality by the use of a two-layer fiber system. Figure 2.7 shows a typical optical fiber arrangement. The core, which is the light containment zone, is surrounded by the cladding, which has a lower refractive index and provides protection to the core surface. This surface is commonly called the core–cladding interface or CCI.

By manufacturing optical fiber in this manner the CCI remains unaffected by external handling or contamination, thereby enabling uninterrupted total internal reflection provided that the light exhibits angles of incidence in excess of the critical angle.

Optical fiber construction and definitions

In the previous section optical fiber was shown to comprise an optical core surrounded by an optical cladding. It is normal convention to define a fiber in terms of its optical core diameter and its optical cladding diameter, measured in microns, where 1 micron equals a thousandth of a millimetre.

Historically a wide range of combinations of core and cladding diameters could be purchased. Over the years rationalization of the offerings has taken place and the generally available formats, known as geometries, are as shown in Table 2.3.

For all the fibers in Table 2.3 the core and cladding are indivisible, i.e. they cannot be separated. This book does not discuss, in detail, the older types of fiber including plastic clad silica, where the cladding was actually removable from the core (with, in some cases, disastrous consequences).

Table 2.3 *Available optical fiber geometries*

Geometry	Core diameter in microns	Cladding diameter in microns	Aspect ratio	Numerical aperture
8/125	8	125	0.064	0.11
50/125	50	125	0.4	0.2
62.5/125	62.5	125	0.5	0.275
100/140	100	140	0.71	0.29

The core and cladding are functionally distinct since:

- The core defines the optical parameters of the fiber (e.g. light acceptance, light loss and bandwidth).
- The cladding is the physical reference surface for all fiber handling processes such as jointing, termination and testing.

Historically the parameter of aspect ratio was used, defined by equation (2.5):

$$\text{aspect ratio} = \frac{\text{core diameter}}{\text{cladding diameter}} \qquad (2.5)$$

The materials used within the core are chosen and manufactured to have higher refractive indices than those of the cladding – otherwise TIR could not be achieved. That being said, there is a variety of processes and materials used to create the core and cladding layers and it will be seen that the difference between the two refractive indices is more relevant to performance than the absolute values.

The ideal fiber

The benefits of optical fiber are shown in Table 2.4. The primary advantages are high bandwidth and low attenuation. The ideal fiber should therefore offer the highest possible bandwidth combined with the lowest possible attenuation. Indeed these two requirements are fulfilled by single mode fiber (8/125). Unfortunately these fibers also accept least light and as a result are difficult use without recourse to expensive injection devices such as semiconductor lasers.

Therefore from the system point of view an ideal fiber does not exist and historically a number of fiber geometries have been developed to meet the needs of particular applications. The following sections discuss the basic fiber parameters of light acceptance, light loss (attenuation or

Table 2.4 *Features and benefits of optical fiber*

Primary	Secondary
Bandwidth – inherently wider bandwidth enables higher data transmission rates over optical fiber leading to lower cable count as compared with copper	*Small size* – fewer cables are necessary leading to reduced duct volume needs
	Light weight – a combination of reduced cable count and material densities results in significant reductions in overall cable harness weight
Attenuation – low optical signal attenuation offers significantly increased inter-repeater distances as compared with copper	*Freedom from electrical interference* – from radio-frequency equipment and power cables
Non-metallic construction – optical fibers manufactured from non-conducting silica have lower material density than that of metallic conductors	*Freedom from crosstalk* – between cables and elimination of earth loops
	Secure transmissions – resulting from non-radiating silica-based medium
	These three factors combine to produce secondary benefits
	Protection – from corrosive environments
	Prevention of propagation of electrical faults – limiting damage to equipment
	Inherent safety – no short-circuit conditions leading to arcing

also known as insertion loss) and bandwidth and attempt to explain the application of different fiber geometries to the diverse environments encountered in telecommunications, military and data communications.

Light acceptance and numerical aperture

The amount of light accepted into a fiber is a critical factor in any cabling design. The calculation and measurement of light acceptance can be complex but its basic concepts are relatively straightforward to understand.

Logically the amount of light accepted into a given fiber must be a function of the quantity of light incident on the surface area of the core for a given light source; otherwise, identical fibers will accept light in direct proportion to their core cross-sectional area. This is defined in equation (2.6):

$$\text{light acceptance} = f \frac{(\pi d)^2}{4} \qquad (2.6)$$

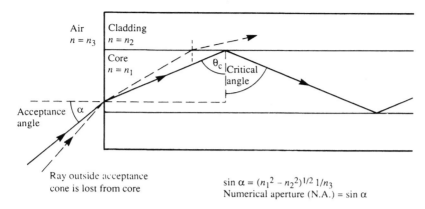

Figure 2.8 *Light acceptance and numerical aperture*

Equally important is the impact of numerical aperture. Referring to Figure 2.8 it can be seen that a ray that meets the first core–cladding interface (CCI) at the critical angle must have been refracted at the point of entry into the fiber core. This ray would have met the fiber core at an angle of incidence (α), which is defined as the acceptance angle of the fiber.

Any rays incident at the fiber core with an angle greater than α will not be refracted sufficiently to undergo TIR at the CCI and therefore, although they will enter the core, they will not be accepted into the fiber for onward transmission.

The term $\sin \alpha$ is commonly defined as the numerical aperture of the fiber and, by reference to Figure 2.8, for $n_3 \approx 1$ (air) then equations (2.7) to (2.10) demonstrate:

$$\sin \alpha \approx (n_1^2 - n_2^2)^{0.5} = \text{numerical aperture (NA)} \quad (2.7)$$

$$\text{light acceptance} = f(\sin \alpha)^2 \quad (2.8)$$

$$f(n_1^2 - n_2^2) \quad (2.9)$$

$$f(\text{NA})^2 \quad (2.10)$$

To summarize, therefore, the amount of light accepted into a fiber is directly proportional to its core cross-sectional area and the square of its numerical aperture.

To maximize the amount of light accepted it is normal to choose fibers with large core diameter and high NA but, as will be seen later in this chapter, these fibers tend to lose most light and have relatively low bandwidths. However, for those environments where short-haul, high-connectivity networks are desirable these fibers are useful and in examples such as aircraft, surface ships and submarines such fibers have found application. In these situations the short-haul requirements minimize the impact of bandwidth and attenuation limitations of fiber geometries with large core diameters and high NA values.

Light loss and attenuation

Transmission of light via total internal reflection has already been discussed and it was stated that no optical power loss takes place at the core–cladding interface. However, light is lost as it travels through the material of the optical core. This loss of transmitted power, commonly called attenuation or insertion loss, occurs for the following reasons:

- intrinsic fiber core attenuation:
  - material absorption;
  - material scattering;
- extrinsic fiber attenuation:
  - microbending;
  - macrobending.

The terms intrinsic and extrinsic relate to the manner in which the loss mechanisms operate. Intrinsic loss mechanisms are those occurring within the core material itself whereas extrinsic attenuation occurs due to non-ideal modifications of the CCI.

Intrinsic loss mechanisms

There are two methods by which transmitted power is attenuated within the core material of an optical fiber. The first is absorption, indicating its removal, and the second is scattering, which suggests its redirection.

Absorption is the term applied to the removal of light by non-reradiating collisions with the atomic structure of the optical core. Essentially the light is absorbed by specific atomic structures which are subsequently energized (or excited) eventually emitting the energy in a different form. The various atomic structures only absorb electromagnetic radiation at particular wavelengths and as a result the attenuation due to absorption is wavelength dependent.

Any core material is composed of a variety of atomic or molecular structures which can undergo excitation thereby removing specific wavelengths of light. These include:

- pure material structures;
- impurity molecules due to non-ideal processes;

- impurity molecules due to intentional modification of pure material structures.

Any pure material has an atomic structure which will absorb selective wavelengths of electromagnetic radiation. It is virtually impossible to manufacture totally pure materials and the absorption spectrum of any impurities serves to modify that of the pure material. Finally in the interest of certain applications it is necessary and desirable to introduce further modifying agents or dopants to improve the performance of the optical core. Most optical fiber is manufactured using a base of pure silica (silicon dioxide – SiO_2) which is doped with germanium and other materials to create an effective core structure. Figure 2.9 shows a typical absorption profile for a silica-based optical fiber.

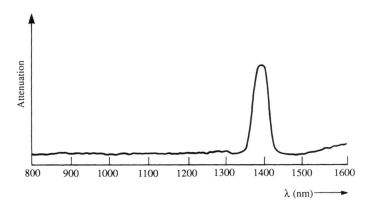

Figure 2.9 *Intrinsic loss characteristic of silica*

Scattering is another major constituent of intrinsic loss. Rayleigh scattering is a phenomenon whereby light is scattered in all directions by minute variations in atomic structure. As shown in Figure 2.10 some of the scattered light will continue to be transmitted in the forward direction, some will be lost into the cladding due the angle of incidence at the CCI being greater than the critical angle and, equally importantly, some will be scattered in the reverse direction via TIR.

Obviously all the light not scattered in the forward direction is effectively lost to the transmission system and scattering therefore acts as an intrinsic loss mechanism.

Rayleigh scattering is wavelength dependent and reduces rapidly as the wavelength of the incident radiation increases. This is shown in Figure 2.11. The absorption spectrum is added to the normal fiber attenuation profile to generate an attenuation profile as shown in Figure 2.12, but it must be

26 Fiber Optic Cabling

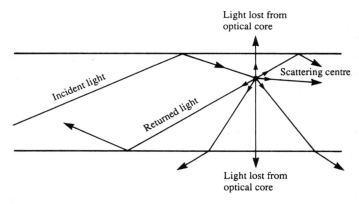

Figure 2.10 Rayleigh scattering in an optical fiber

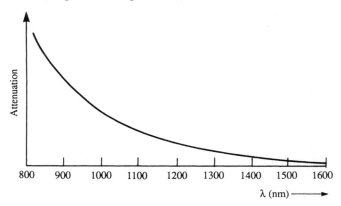

Figure 2.11 Rayleigh scattering characteristic of silica

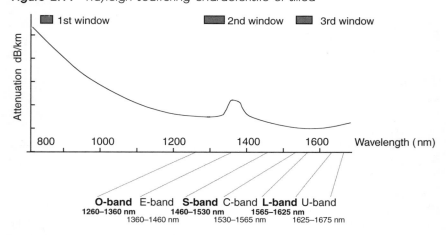

Figure 2.12 Typical attenuation profile of an optical fiber

pointed out that the above intrinsic loss mechanisms are linked to the length of the light path within the fiber and not to length of the fiber itself.

Modal distribution and fiber attenuation

The numerical aperture (NA) of an optical fiber is a measure of the acceptance angle of that fiber which, in turn, is related to the critical angle of that fiber by formulae (2.11) and (2.12):

$$\text{NA} = \frac{n_2}{\tan \theta_c} \tag{2.11}$$

$$\theta_c = \tan^{-1}(n_2/\text{NA}) \tag{2.12}$$

Light may take many different paths along the optical core ranging from the 'zero-order mode' to the 'highest-order mode' as shown in Figure 2.13. Interestingly, an infinite number of modes is not possible as will be seen later in this chapter. Rather a given fiber can only support a specific number of transmitting modes which is a complex combination of core diameter (d), wavelength of transmitted light and the NA of the fiber.

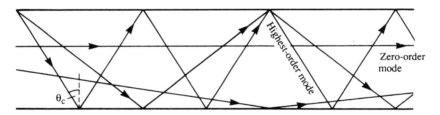

Figure 2.13 *Transmission modes within the optical core*

A statistical analysis will show that all the modes are not equally populated and that only a small proportion of the total light is transmitted by highest-order modes. Nevertheless the higher NA fibers will contain light travelling at higher angles of incidence than those with lower NA values. Higher-order modes incur greater path lengths for a given length of fiber as the calculations in Table 2.5 show. Therefore fibers which can support higher-order modes will necessarily exhibit higher levels of loss due to intrinsic attenuation mechanisms.

Therefore fibers that accept most light will also lose most light. As a result fibers with low NA are necessary for long-range communication. Figure 2.14 shows the typical attenuation profiles for a number of fibers and clearly demonstrates the impact of numerical aperture on attenuation.

Table 2.5 *Maximum optical path lengths for stepped index constructions*

Numerical aperture	Acceptance angle (α) (degrees)	Critical angle θ_c (degrees)	Maximum optical path length $\left(\dfrac{L_{max} - L}{L}\right)\%$
0.11	6.32	85.74	0.28
0.20	11.54	82.25	0.92
0.26	15.07	79.90	1.57
0.275	15.96	79.32	1.76
0.29	16.86	78.73	1.97
0.45	26.74	72.35	4.93

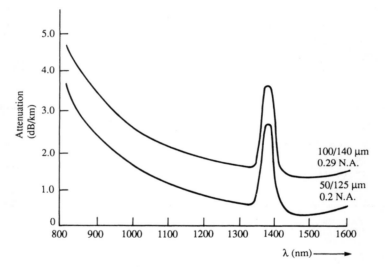

Figure 2.14 *Attenuation variations with optical fiber geometry*

Extrinsic loss mechanisms

Extrinsic losses are those generated from outside the confines of the optical core which subsequently affect the transmission of light within the core by damaging or otherwise modifying the behaviour of the CCI.

These loss mechanisms are generally of two types:

- *Macrobending.* Light lost from the optical core due to macroscopic effects such as tight bends being induced in the fiber itself.

Optical fiber theory 29

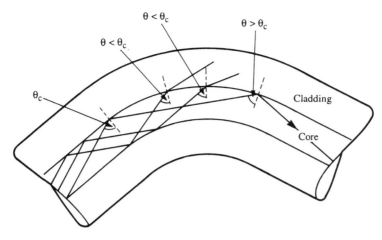

Figure 2.15 *Removal of light from optical core via macrobending*

- *Microbending.* Light lost from the optical core due to microscopic effects resulting from deformation and damage to the CCI.

Severe bending of an optical fiber is depicted in Figure 2.15 and it is clear that light can be lost when the angle of incidence exceeds the critical angle. This type of macrobending is common but is obviously more pronounced when fibers with low NA are used, since the critical angles are larger.

Macrobending losses are normally produced by poor handling of fiber rather than problems in fiber or cable manufacture. Poor reeling (see Figure 2.16) and mishandling during installation can create severe bending of the fiber resulting in small but important localized losses. These are

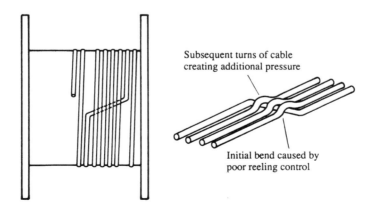

Figure 2.16 *Macrobending due to poor reeling*

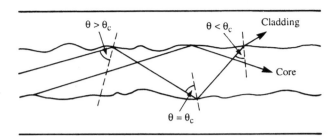

Figure 2.17 *Removal of light from the optical core due to microbending*

unlikely to prevent system operation but they are indicative of significant stresses occurring at the cladding surface which will have potential lifetime limitations.

Microbending is a much more critical feature and can be a major cause of cabling attenuation. It is normally seen where the CCI is not a smooth cylindrical surface. Rather, due to processing or environmental factors, it becomes modified or damaged as is shown in Figure 2.17. Such defects can result from the generation of compressive and shear stresses at the CCI, creating a rippling effect which can radically alter the transmission of light. These stresses are very difficult to define; however, they can be caused by:

- poor fiber processing;
- incorrect processing during cabling;
- low temperatures;
- high pressures.

Microbending may be present and widespread throughout a fiber which can be detected as attenuation which exceeds specification. Often, however, the fiber may appear normal under standard conditions but may exhibit severe microbending losses when handled. These are seen during the application of a connector where external force (which would have no effect on a normal fiber) can create significant losses in fibers subject to microbending tendencies.

Low temperatures and high pressures can have similar effects and careful design is necessary for cables required to function in these environments. As with macrobending, fibers with low numerical apertures are more easily affected for a given change in the CCI angle. (This should be easily understood since an equal modal distribution of power suggests that if β_c = 80° (NA approx. 0.26) then a CCI modification of 5° may influence as much as (5/10) 50% of the transmitted power whereas if θ_c = 70° (NA = 0.50) then only 25% will be affected. Modal distribution may not reflect these theoretical results; however, it does allow a basic understanding to be gained.)

It should also be noted that microbending losses can be very sensitive to wavelength and, in general, losses become more severe as wavelengths increase. This is discussed later in this chapter.

Impact of numerical aperture on attenuation

It was stated earlier in this chapter that to maximize light injection it is desirable to adopt a fiber geometry with large core diameter and high NA.

This section has shown that this fiber design will generally exhibit greater attenuation than that exhibited by a geometry with low NA. The final trade-off appears in the form of extrinsic loss mechanisms being less dominant in fibers with high NA.

A complex set of options exists but market-led rationalization has resulted in a range of fiber geometries to suit telecommunications and data communications with a further specialist group to meet the requirements of the military environment.

The first set exhibit core diameters between 8 and 100 microns with NA values between 0.11 and 0.29, whereas fibers for high-connectivity and high-pressure environments may have 200 micron diameter cores and NA values of 0.4 and above.

Operational wavelength windows

Figure 2.12 shows a typical attenuation profile for a silica-based fiber. Three operational wavelength windows are highlighted which have long been established as the bases for fiber optic data transmission.

The first window, centred around 850 nm, was the original operating wavelength for the early telecommunications systems and is now the main system wavelength for data and military communications systems.

The second window at 1300 nm evolved for high-speed telecommunications both because of reduced levels of attenuation and, as will be discussed later in this chapter, because of the low levels of intramodal dispersion. Its bandwidth performance has led to its adoption for the higher-speed data communications standards such as FDDI (fiber distributed data interface) and Fast Ethernet 100BASE-FX.

The third window at 1550 nm is used for long distance telecommunications and has been proposed in a data communication standard for the first time, i.e. the 10GBASE-EW proposal for gigabit Ethernet.

Bandwidth

The bandwidth of optical fiber was one of the primary benefits leading to its adoption as a transmission medium for telecommunications. Indeed, once the attenuation of fibers had made possible minimal configurations of repeaters/regenerators, improvements in operational bandwidth were the prime motivation behind many technological developments.

An understanding of the bandwidth limitations of optical fiber is necessary to enable the network designer and installer alike to assess the correctness of a particular component choice.

Optical fiber and bandwidth

The prerequisite of any communications system is that data injected at the transmitter shall be detected at the receiver in the same order and in an unambiguous fashion. The impact of any defined, limited bandwidth is that, at some point, data becomes disordered in the time domain.

With regard to optical fiber this results from simultaneously transmitted information being received at different times. The mechanisms by which this may occur are:

- intermodal dispersion;
- intramodal (or chromatic) dispersion.

Dispersion is a measure of the spreading of an injected light pulse and is normally measured in seconds per kilometre or, more appropriately, picoseconds per kilometre.

Intermodal dispersion

Earlier in this chapter it was explained that light travels in a defined and limited number of modes, N, where equation (2.13) gives:

$$N = \frac{0.5(\pi d(\mathrm{NA}))^2}{\lambda^2} \tag{2.13}$$

These modes range from the highest-order mode, which comprises light travelling at the critical angle, to the zero-order mode which travels parallel to the central axis of the fiber.

The term 'intermodal dispersion' relates to the differential path length (and therefore transmission time) between the highest-order mode and the zero-order mode.

Figure 2.18 details the relevant equations for path length which show the differential path length and time dispersion due to intermodal dispersion, i.e. dispersion due to timing differences between modes.

Using equation (2.14):

$$\Delta T = 333(n_1 - n_2) \text{ ns/km} \tag{2.14}$$

it can be seen that for $n_1 = 1.484$

Optical fiber theory

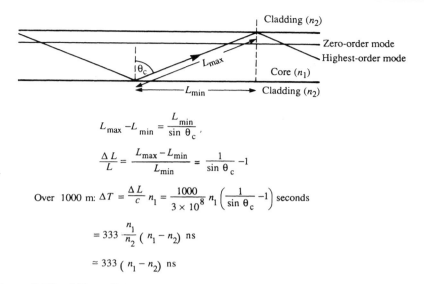

Figure 2.18 *Differential path length and timing*

NA = 0.2 ΔT = 5 ns/km
NA = 0.5 ΔT = 29 ns/km

The relationship between dispersion and bandwidth is very much based upon a practical approach but is generally accepted to be as stated in equation (2.15):

$$\text{Bandwidth (Hz.km)} = \frac{3.1}{\Delta T \text{ (s/km)}} \qquad (2.15)$$

Which suggests that no more than about three individual short pulses may be injected within a given dispersion period, otherwise the received data will be disordered to the point of unacceptability.

This equation is more frequently seen as in (2.16):

$$B = \text{bandwidth (MHz.km)} = \frac{310}{\Delta T \text{ (ns/km)}} \qquad (2.16)$$

Which with regard to the above fiber geometries suggests that

NA = 0.2   B = 67 MHz.km
NA = 0.5   B = 11 MHz.km

Performance figures such as these represent a significant shortfall against the enormous bandwidths promised by optical fiber technology. As the primary technical mission was for improvements in bandwidth then the refinement of processing techniques was totally dominated by developments in this area. This refinement led to the production of graded index fibers.

Step index and graded index fibers

So far in the treatment of optical fiber it has been assumed that the core and cladding comprise a two-level refractive index structure: a high-index core surrounded by cladding with a lower refractive index. This type of fiber is defined as stepped or step index because the refractive index profile is in the form of a step (see Figure 2.19). This design of fiber construction is still used for large core diameter, large NA, fibers as discussed earlier in this chapter where applications required a high level of light acceptance.

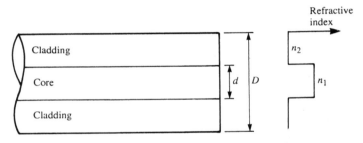

Figure 2.19 *Stepped refractive index profile*

As an alternative, the concept of a graded index fiber was considered.

Graded index optical fibers are manufactured in a comparatively complex manner and they feature an optical core in which the refractive index varies in a controlled way. The refractive index at the central axis of the core is made higher than that of the material at the outside of the core. The effect of the lower refractive index layers is to accelerate the light as it passes through.

The higher-order modes, which spend proportionally more time away from the centre of the core, are therefore speeded up with the intention of narrowing the time dispersion and hence increasing the operating bandwidth of the fiber. The refractive index layers are built up concentrically during the manufacturing process. Careful design of this profile enabled not only the acceleration effect mentioned above but also had a secondary but no less important function. Treating the profile as shown in Figure 2.20 it is clear that the light no longer travels as a straight line but is curved, although at the microscopic level this curve is composed of a series of straight lines.

Optical fiber theory 35

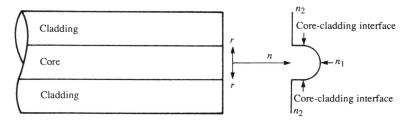

Core refractive index profile

$$n_r = n_1 \left(1 - \frac{n_1^2 - n_2^2}{n_2^2} \cdot \frac{r^2}{d^2}\right)$$

Figure 2.20 *Graded refractive index profile*

The ideal graded index profile balances the additional path lengths of higher-order modes with the increased speeds of travel within those modes, thereby reducing intermodal dispersion considerably. This was achieved by producing a profile defined by equation (2.17):

$$n(r) = n_1 \left(1 - \frac{2s(r^2)}{d^2}\right) \tag{2.17}$$

Where s is given by (2.18):

$$s = (n_1^2 - n_2^2)/n_2^2 \tag{2.18}$$

Increases in bandwidths of the order of tenfold were achieved and at this stage it was felt that optical fiber had reached the point of acceptability as a carrier of high-speed data.

Manufacturing tolerances have continually been improved and profiles have been more closely controlled. As a result operating bandwidths of graded index fibers have gradually increased to the point where little further improvement is possible, although the manufacturing process can be tuned and product selected to give some outstanding performances in excess of 1000 MHz.km.

The majority of optical-based data communications networks today utilize graded index (GI) profiles. However, 97% of the world market for optical fiber (year 2000) is for single mode, sometimes known as monomode. Single mode fiber is addressed in more detail later in this chapter.

The use of GI profiles modifies the equations detailed earlier in this chapter. In particular equation (2.7):

$$(n_1^2 - n_2^2)^{0.5} \tag{2.7}$$

A graded index fiber with an ideal profile will have a modified NA as shown in (2.19):

$$(0.5(n_1^2 - n_2^2))^{0.5} \approx \text{numerical aperture, graded index} \qquad (2.19)$$

It can be seen then that the NA of a graded profile fiber will be lower than that of an equivalent stepped index fiber. Once again it is logical to expect large core diameter, high NA (high light acceptance) fibers to feature stepped index core structures whereas the higher-bandwidth, low attenuation fibers will feature smaller core diameters and will utilize as low an NA value as possible; normally achieved by the use of a graded index core structure.

Modal conversion and its effect upon bandwidth

In the previous sections we have treated the fiber core as being able to support a large but finite number of transmission modes ranging from the zero-order mode (travelling parallel to the axis of the core) to the highest-order mode (travelling along the fiber at the critical angle). This assumption allows the calculation of attenuation and bandwidth dependencies as has already been shown. Needless to say this ideal model is far from the truth. It is unlikely, not to say impossible, to manufacture and cable a fiber such that propagation of the modes would continue in such an orderly fashion.

In practice there is a continual process of modal conversion (see Figure 2.21) changing zero-order modes to higher orders and vice versa. In any fiber it is fair to assume that bandwidth performances will exceed theoretical models outlined here, be they of stepped or graded index profile, due to the equalization of path lengths.

The practical improvements in bandwidths due to modal conversion may be as much as 40%, which pushed fiber bandwidths into the region of 1000 MHz.km. However, these levels were not sufficient for the telecommunications market with their huge cabling infrastructure requiring

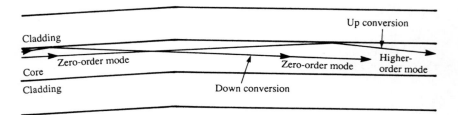

Figure 2.21 *Modal conversion*

to be as 'future-proof' as possible. One further technical leap was necessary: single mode (or monomode) optical fiber.

Differing multimode bandwidths when using lasers and LEDs

Up until around 1996 multimode fiber was generally used with light emitting diodes (LEDs) as the transmitting devices. LEDs are low cost and their ability to send up to 200 Mb/s over 2 kilometres of multimode fiber was seen as more than adequate. The bandwidth of multimode fiber was measured using the output conditions of an LED which uniformly illuminates the core of the fiber and excites hundreds of modes in the core. This is known as overfilled launch (OFL) and the resulting bandwidth, when measured with such an input device, is known as overfilled launch bandwidth (OFL-BW). This is shown in Figure 2.22.

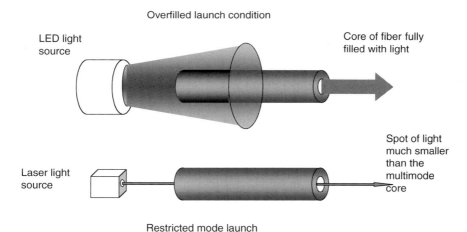

Figure 2.22 *Overfilled and restricted mode launch conditions*

Unfortunately LEDs can only be modulated up to a few hundreds of megahertz, and so are not generally suitable for gigabit speed transmission. In November 2000 a Japanese laboratory demonstrated a 1 Gb/s superluminescent InGaAs/AlGaAs LED operating at 880 and 930 nm. This device, however, remains in the lab for the present and the vast majority of gigabit transmission equipment will need lasers to achieve gigabit speeds.

The problem with lasers is that traditionally they are very expensive, having been developed with long-haul telecommunications in mind. The two styles of laser used in single mode telecommunications systems are known as distributed feedback (DFB), and Fabry–Perot (FP). The whole rationale of using multimode optical fiber is that it allows the use of cheap LEDs (an LED cannot focus enough power into the core of a single mode fiber) even though single mode fiber, with its higher bandwidth and lower attenuation, is around one-third the price of multimode.

In the mid-1990s a new style of laser was developed, known as a VCSEL, or vertical cavity surface emitting laser. This effectively is a laser with the manufacturing costs of a good LED. VCSELs are currently available at 850 nm and can inject gigabit datastreams into multimode fiber. VCSEL 850 nm transmission devices are assumed for use in the optical gigabit Ethernet standard 1000BASE-SX (IEEE 802.3z) which was published in 1998: 850 nm VCSELs will also be used in Fiber Channel, ATM and other Ethernet physical layer presentations. A single mode 1300 nm VCSEL is currently working its way through the laboratories of the world, a device that could totally transform the economics of using single mode fiber in campus and premises cabling.

The problem has arisen though that the apparent bandwidth of an overfilled launch multimode fiber is not necessarily the same as when that same fiber is used with a laser.

An LED evenly illuminates the whole core of the fiber whereas a laser sends a spot of light much smaller than the core diameter. Laser launch should give a higher bandwidth than OFL because fewer modes are being excited therefore less modal dispersion should occur. This would only be true, however, if the graded index profile of the multimode fiber were near perfect, but this is not always the case. Small defects, especially at the very centre of the core, change the refractive index, and hence the speed of the lower order modes, in an unpredictable way. This effect is known as differential mode delay (DMD) and reduces the available bandwidth of the optical fiber. This effect has to be taken into account when using gigabit Ethernet (IEEE 802.3z) and in certain equipment requires the use of a 'mode-conditioning cord'. This special equipment cord splices a single mode fiber from the transmitter onto a multimode fiber which is then onward connected to the multimode cabling. Normally, when splicing fibers together, the engineer would attempt to get exact core-to-core alignment, but with the offset lead a deliberate 17 to 23 micron offset is introduced so that the laser spot from the transmitter is not being launched immediately into the core of the 62.5/125 multimode fiber (for 50/125 fiber the offset is 10 to 16 microns).

Differential mode delay and uncertainties over laser launch bandwidth will be of even more importance when using 10 000BASE-SX, i.e. ten gigabit Ethernet over multimode fiber.

The American Telecommunications Industry Association (TIA) formed FO-2.2.1 Task Group on Modal Dependence of Bandwidth to assess measurement techniques for both OFL and laser launch bandwidth. This committee concluded that two criteria need to be understood:

Optical fiber theory 39

- Transceivers must meet source encircled flux requirement (EF).
- The fiber needs to meet a requirement for restricted mode launch (RML) criterion.

EF is the percentage of power within a given radius when a transmitter launches light into a multimode fiber. The EF measurement characterizes the spot size carried by the multimode fiber and assists in consistent launch conditions between different manufacturers of VCSELs.

The RML bandwidth measurement launch condition is created by filtering the overfilled launch condition with a special 23.5 micron fiber. This RML fiber must be at least 1.5 metres long to eliminate leaky modes and less than 5 metres long to avoid transient loss effects.[1] A new term to be introduced is effective modal bandwidth, EMB, which measures the effects of multimode fiber and transceiver interaction to accurately evaluate system performance.

The different bandwidth performance of multimode fiber under different launch conditions has been recognized by ISO 11801 2nd Edition. Three performance grades of multimode fiber are specified, OM1, OM2 and OM3. The bandwidth requirements are detailed in Table 2.6. OM3 fiber is specifically designed as laser enhanced or optimized. OM3 is 50/125 fiber with a very large bandwidth at the first window, such as 2000 MHz.km, and a near perfect refractive index profile. This kind of fiber will be needed for the 850 nm multimode version of ten gigabit Ethernet, 10 000BASE-SX, and is expected to provide extended distances for gigabit Ethernet transmission.

Table 2.6 *ISO 11801 optical fiber modal bandwidth*

Optical fiber class	Performance utilizing LED sources	Performance utilizing laser sources
OM1	up to 2000 m and up to 155 Mb/s	up to 300 m and up to 1Gb/s
OM2	up to 2000 m and up to 155 Mb/s	up to 500 m and up to 1 Gb/s
OM3	up to 2000 m and up to 155 Mb/s	up to 300 m and up to 10 Gb/s
OS1	not applicable	up to 2000 m (or more) and up to 10 Gb/s or more

Single mode transmission in optical fiber

The relatively high bandwidths indicated above were achieved in the late 1970s but they fell short of the requirements of the telecommunications markets and large-scale applications could not be envisaged without the introduction of next-generation fiber geometries.

Intermodal dispersion occurs since the time taken to travel between the two ends of a fiber varies between the transmitted modes. Graded index profiles served to reduce the effects of this phenomenon; however, manufacturing variances limited the maximum achievable bandwidths to a level considered restrictive by the long-haul market where data rates in excess of 1 gigabit/s were forecast with unrepeated links of 50 km being considered.

The logical approach to elimination of intermodal dispersion was simply to eliminate all modes except for one. Thus the concept of single or monomode optical fiber was born.

With reference to equation (2.13) it is seen that the number of modes within a fiber are limited to defined, discrete values of N as shown in equations (2.20a) and (2.20b):

$$N = \frac{0.5(\pi d(\mathrm{NA}))^2}{\lambda^2} \quad \text{for stepped index fibers, and} \qquad (2.20a)$$

$$N = \frac{0.25(\pi d(\mathrm{NA}))^2}{\lambda^2} \quad \text{for graded index fibers} \qquad (2.20b)$$

To derive this formula it is necessary to modify the mathematical treatment of light from that of a ray (as in the early part of this chapter for reflection and refraction) and a photon (as in the consideration of absorption by atomic or molecular excitation) to that of a wave.

As the wavelength of transmitted electromagnetic radiation increases, the number of modes that can be supported will fall. Extended to its limit this applies to microwave waveguides. Similarly as core diameter (d) and NA fall so does the value of N.

As will be seen later the optimum wavelength to be adopted for long-haul, high-bandwidth transmission is 1300 nm, although one can specify dispersion shifted single mode fiber optimized for 1550 nm and very long-haul operation.

A stepped index fiber was created with an NA value of approximately 0.11 and core diameter of 8 μm. The insertion of these parameters into this equation suggests that N is less than 2. As the number of modes must be an integer then this means that only a single transmitted mode is possible and that this light will travel as a wave within a waveguide independent of the nature of the light before it entered the fiber.

A critical diameter and cut-off wavelength are defined in equations (2.21) and (2.22):

$$d_c = \frac{2.4\lambda}{\pi \mathrm{NA}} \qquad (2.21)$$

and

$$\lambda_c = \frac{\pi d(\text{NA})}{2.4} \tag{2.22}$$

If the core diameter falls below, or the wavelength rises above, these figures then the fibers act as single mode media, otherwise the fibers operate as multimode elements.

This is a difficult concept to come to terms with and in most cases its blind acceptance is adequate. Those readers wishing to become more deeply involved are referred to standard wave optics texts which, although highly mathematical, should provide some degree of satisfaction.

Single mode or monomode optical fiber therefore exhibits an infinite bandwidth due to zero intermodal dispersion. In practice of course the single mode fibers in common use do not have infinite bandwidths and are limited by second-order effects such as chromatic and polarization mode dispersion, which are discussed later in this chapter. Nevertheless the bandwidths achieved by single mode optical fibers are extremely high and the single mode technology has been adopted universally for telecommunications systems worldwide.

Furthermore it should be noted that if only one mode can be propagated by the fiber due to the mathematics of standard wave theory then there is no need for a graded index profile within the core itself. As a result the fibers are significantly cheaper to produce, a factor which is further enhanced by the low NA value (requiring lower levels of dopant materials). Additionally the low NA provides the single mode fiber with a significantly lower attenuation performance than its multimode counterparts.

Single mode technology therefore provides the communications world with a remarkable paradox. The highest-bandwidth, lowest-attenuation, optical fiber is the cheapest to produce. This factor is only just starting to impact the data communications markets but the trend is undeniably towards single mode technology in all application areas.

Attenuation and single mode fiber

As indicated above a given optical fiber may operate as either single mode or multimode, dependent on the operating wavelength. Equation (2.22) suggests that as transmission wavelengths decrease a single mode fiber may become multimode as the number of modes, N, increases to 2 and above.

This transformation has consequences for the attenuation of the optical fiber as can be seen from Figure 2.23, which shows the two attenuation profiles. Multimode transmission results in higher attenuation than that experienced in the same fiber under single mode operation.

The wavelength at which this change takes place is called the cut-off wavelength λ_c, as shown in equation (2.23):

$$\lambda_c = \frac{\pi d(\mathrm{NA})}{2.4} = 1.306 d(\mathrm{NA}) \qquad (2.23)$$

Typical values of cut-off wavelength lie between 1260 nm and 1350 nm. Note that a fiber datasheet may choose rather to specify cable cutoff wavelength, λ_{cc}, as a more useful figure to the end user.

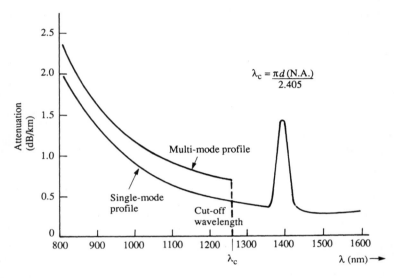

Figure 2.23 *Multimode – single mode attenuation profiles*

Mode field diameter

Within a fiber of 8 μm core diameter, light will travel as a wave. The wave is not fully confined by the core–cladding interface and behaves as if light travels partially within the cladding, and as a result the effective diameter is rather greater than that of the core itself.

Mathematically a mode field diameter is defined in (2.24):

$$\mathrm{MFD} = \frac{2.6\lambda}{\pi(\mathrm{NA})} \qquad (2.24)$$

This parameter is relevant in the calculation of losses and parametric mismatches between fibers. Typical mode field diameters are 9.2 microns at 1310 nm and 10.4 microns at 1550 nm.

Intramodal (or chromatic) dispersion

Intramodal dispersion is a second-order, but none the less important, effect caused by the dependency of refractive index upon the wavelength of electromagnetic radiation. This was mentioned earlier in this chapter and referring back to Table 2.2 it can be seen that for typical fiber materials the refractive index varies significantly across the range of wavelengths used for data transmission. In the graph, at 850 nm, it will be noticed that the rate of change of refractive index is as severe as it is at 1550 nm. In comparison the dependence is much reduced at 1300 nm. The relevance of this phenomenon is that when linked with the spectral width of light sources, another mechanism for pulse broadening is seen which in turn limits the potential bandwidth of the fiber transmission system.

At this point the detailed explanation of refractive index must be taken a little further.

Earlier in this chapter it was stated that the refractive index of all dispersive materials must be greater than unity since it is well established that the speed of light in a vacuum is a boundary which cannot be crossed (except in science fiction). However, it is possible for a material to have a refractive index lower than unity for a single wavelength signal. Unfortunately, such a signal can carry no information and therefore its velocity is irrelevant.

Real signals depend upon combinations of single or multiple wavelength signals. These combinations may be in the form of pulses. The information content within these signals travels at a speed defined by the group refractive index which is related to the spectral refractive index as shown in equation (2.25):

$$n_g = \frac{n}{\left(1 - \frac{\lambda}{n} \cdot \frac{dn}{d\lambda}\right)} \qquad (2.25)$$

As $dn/d\lambda$ is always greater than unity then the group refractive index n_g is always greater than n. Figure 2.25 shows the group refractive index profile for pure silica and indicates a minimum at 1300 nm.

If light emitting diodes (LEDs) are compared with devices which generate light which is amplified due to stimulated emission of radiation (lasers) it can be seen that the range of discrete wavelengths produced is significantly lower for the laser devices. Figure 2.24 shows a typical spectral distribution for the two device types. Because the group refractive index varies across the spectral width of the devices the resulting light will travel at different speeds. The differential between these speeds is the source of pulse broadening due to intramodal or chromatic dispersion. Based upon the information in Table 2.2 and Figure 2.25 then it is clear that the dispersion will be lowest for a laser operating around the second window and highest for an LED at 850 nm.

Therefore bandwidth of fiber is not always a straightforward issue of the optical fiber itself but also depends upon the type of devices used to transmit the optically encoded information.

44 Fiber Optic Cabling

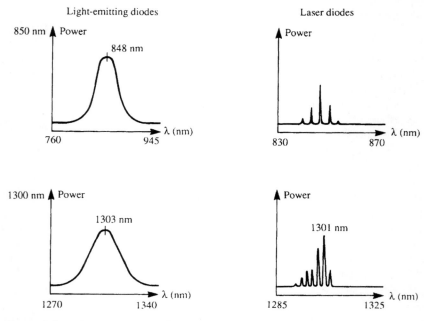

Figure 2.24 Spectral distribution of transmission devices

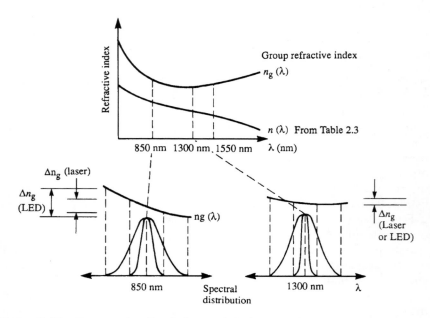

Figure 2.25 Group refractive index and device spectra

Polarization mode dispersion

Polarization mode dispersion or PMD, is a further type of bandwidth limiting dispersion that affects single mode fiber when operating at long distances and/or high bit rates. Single mode fiber supports two orthogonal polarizations of the original transmitted signal. Random variations in the refractive index along with non-circularity of the core leads to the two polarized modes travelling with different velocities and hence spreading out in time. The effect is length related and will affect high bit rate signals more as the time frame between bits will be subsequently shorter and therefore more susceptible to interference from the trailing polarized mode. Analogue video signals will also suffer interference in the same way.

The units of PMD are picoseconds over the square root of the distance (in kilometres). Some manufacturers quote their optical cable fiber performance as having a worst case value of 0.5 $ps.km^{-1/2}$, whilst other proposals have been made to quote a maximum uncabled-fiber performance of around 0.2 to 0.3 $ps.km^{-1/2}$. Another term used in some datasheets is the PMD link value, which is a statistical value of PMD over a number of concatenated lengths of fiber. This figure is thought to give a more realistic idea of system performance rather than the worst case figures. A typical PMD link value is < 0.1 $ps.km^{-1/2}$.

Bandwidth specifications for optical fiber

The impact of both intermodal and intramodal dispersion upon the bandwidth of optical fiber is quite clearly seen in component specifications.

For intermodal dispersion the key parameter is numerical aperture, and the dependence is clearly seen in Table 2.7. For 50/125 micron 0.2 NA

Table 2.7 *Bandwidth performance values*

Fiber geometry	NA	Potential bandwidth (MHz.km)	Typical commercial specification for optical fiber bandwidth (MHz.km)		
			Laser 1300 nm	LED 850 nm	LED 1300 nm
8/125	0.11	infinite	>10 000	—	—
50/125	0.20	2000		500	500
62.5/125	0.275	1000		160	500
100/140	0.29	500		100	250

fiber the first window overfilled launch (850 nm LED) bandwidth is typically 500 MHz.km, whereas the equivalent figure for a 62.5/125 micron 0.275 NA fiber is 160 MHz.km.

The physical removal of modes and the low 0.11 NA achieved by single mode fibers extends this bandwidth; however, the quoted figures are not infinite. This restriction is primarily due to intramodal dispersion which also affects the larger core fibers.

The danger comes in accepting the figures too readily and at face value. If an LED could be driven at the same high transmission rates normally attributed to lasers (in excess of 500 Mb/s) there is no guarantee that the communication system would operate correctly because the available bandwidth would be considerably lower than the potential bandwidth.

The concept of available bandwidth (which allows for the transmission equipment) as opposed to the potential bandwidth (as measured by narrow-spectrum devices) can be important, particularly at the design stage.

System design, bandwidth utilization and fiber geometries

Returning to a comment made early in this chapter where it was stated that the two main reasons for non-operation of an optical fiber data highway are, first, a lack of optical power and, second, a lack of available bandwidth.

Telecommunications systems require very high bandwidths and low losses. This need is serviced by single mode fiber, laser transmission equipment and second window operation. The latter is obviously no accident as it combines a low attenuation with the lowest levels of intramodal dispersion.

Lasers produce light in a low NA configuration enabling effective launch into the 8 micron optical core, whereas LED sources tend to emit light with a more spread-out, higher NA content. Accordingly their use has required a larger core diameter, high NA fiber design to produce acceptable launch conditions. These fibers exhibit lower intermodal bandwidths and when combined with the wider spectrum of the LED also feature a lower level of intramodal bandwidth. The large core diameter, high NA fibers tend to exhibit higher levels of optical attenuation, which supports their use over short and medium distances, which in turn removes the need for the tremendously high bandwidths of the single mode systems.

Transmission distance and bandwidth

The variation of bandwidth with distance is not a simple relationship. The mechanisms by which dispersion takes place are certainly distance related and logically it can be assumed that a linear relationship exists.

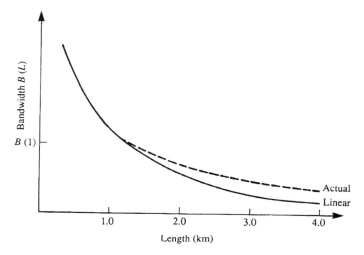

Figure 2.26 *Bandwidth and system length*

This means that 500 metres of 500 MHz.km optical fiber will have a bandwidth of approximately 1000 MHz. However, tests have shown that long lengths of optical fiber exhibit bandwidths better than linear. This effect is thought to be due to changes of modal distribution at connectors and/or spliced joints etc. This modal mixing is in addition to the modal conversion already discussed and serves to further reduce the intermodal dispersion.

Figure 2.26 shows the probable variation of bandwidth with distance; however, accurate measurement is difficult in the field and the measured value can vary following reconnection of the system due to differing degrees of modal mixing. It is safest therefore to assume a linear relationship for all distances particularly in the data communications field where interequipment distances are rarely longer than 2 kilometres.

Optical fiber geometries

Currently there are three mainstream fiber geometries: 8/125 micron for telecommunications, 50/125 micron and 62.5/125 micron for data communications. Single mode fiber has now extended into the data communications/LAN market to the extent that for ten gigabit Ethernet, any transmission distance beyond 300 metres requires single mode.

The market has not always been so standardized and in the early days of fiber many different styles could be purchased – almost to be made to order. This was particularly true of the early military market, where special requirements frequently produced fiber styles purely for a single operational need. Usage was not significant, prices were high and in that environment it was feasible to have exactly what was required rather than accept any view of standardization.

As a result many geometries have existed and Table 2.8 is not exhaustive.

Table 2.8 *Some fiber geometries and constructions*

Fiber	Refractive Index Profile	Construction
8/125	stepped	doped silica
35/125	stepped	doped silica
50/125	graded	doped silica
62.5/125	graded	doped silica
85/125	stepped	doped silica
85/125	graded	doped silica
100/140	graded	doped silica
200/230	stepped	clad silica
200/280	stepped	plastic clad silica
980/1000	stepped	plastic

Single mode, 50/125 and 62.5/125 represent 99% of the optical fiber market nowadays, with large core fibers such as 200/280 hard clad silica and all-plastic fibers being used mainly for short-distance, low-bandwidth instrumentation links.

62.5/125 optical fiber has been standardized in the American, European and ISO standards since the mid-1990s, and is the predominant fiber type used in premises cabling in most countries. 50/125 is recognized in the European standards (EN 50173) and ISO 11801 and has always been the market leader for LAN optical cabling in Germany, Japan and South Africa. 62.5/125 fiber is starting to show its age and limitations. The advent of gigabit LANs such as gigabit and ten gigabit Ethernet have shown that link lengths can be reduced to as low as 220 metres for gigabit Ethernet and less than 50 metres for ten gigabit Ethernet when using the most common grade of 62.5/125. 50/125 has started to make a comeback because it has a higher bandwidth, lower attenuation and generally costs less than 62.5/125. New, laser optimized, multimode fibers are generally based on 50/125 and 50/125 has reappeared in the American standard TIA/EIA 568B.

The new family of single mode fiber

Standard single mode fiber, which was developed in the early 1980s, was optimized for use at around 1310 nm, principally meaning that the chromatic zero dispersion point is placed around this wavelength. Standard 1310 single mode fiber, sometimes referred to as SMF, is still the principal fiber in use today, but the advent of optical amplifiers and the need to go longer distances encouraged researchers to study 1550 nm performance. 1550 nm operation has a lower attenuation, typically 0.25 dB compared to 0.35 at 1300 nm, but the dispersion is much worse, e.g. 18 ps/nm.km.

Single mode fiber is described in ITU-T Recommendation G.652 and IEC 60793-2. A typical fiber is Corning SMF28 which has a zero dispersion wavelength between 1301 and 1325 nm.

Optical amplifiers, known as EDFAs, erbium doped fiber amplifiers, work at 1550 nm and require an optical fiber with dispersion minimized at 1550 nm. In the late 1980s dispersion shifted fiber (DSF) was introduced which has the zero dispersion point centred at 1550 nm.

The optical spectrum as applicable to optical fiber is now subdivided into bands as well as operating windows. For example, C-Band is 1530 to 1565 nm and L-band is 1565 to 1625 nm. S-band is 1450 and 1500 nm. The water absorption peak, which all fibers suffer from, is at 1383 ± 3 nm, and transmission equipment will always try and keep clear of this zone.

Increasing optical output powers revealed a problem with single mode fiber, especially DSF, based on a non-linear performance. At high power, the fiber exhibits a non-linear response, and in any system, electronic or optical, a non-linear response leads to a number of intermodulation effects. For DSF optical fibers these can be summarized as self-phase modulation, cross-phase modulation, modulation instability and four-wave mixing. Of these, four-wave mixing is regarded as the most troublesome. Four-wave mixing generates a number of 'ghost channels' which not only drain the main optical signal but also interfere with other optical channels. For wavelength division multiplexing, WDM, this is a severe problem, for example a 32 channel WDM has a potential 15 872 mixing components. Four-wave mixing is most efficient at the zero dispersion point.

A new fiber was therefore introduced called non-zero dispersion shifted fiber, NZ-DSF. NZ-DSF has the zero dispersion point moved up to around 1566 nm. This is just outside the gain band of an EDFA. So a small amount of dispersion is introduced into the system but at a controlled amount and with minimized four-wave mixing. NZ-DSF is described in ITU-T Recommendation G.655.

For the vast installed base of standard single mode, it is possible to add quantities of dispersion compensating fiber, DCF, to equalize the dispersion effects. DCF has an opposite dispersive effect to 1310 single mode and can reverse pulse spreading to some extent.

Another approach is to make the core area of the NZ-DSF larger. A larger core will have a lower energy density and so four-wave mixing effects will be much lower. So-called large effective area fibers have a core area around 30% larger than conventional single mode fibers and this gives a much lower energy density. There has to be a limit of course. Making the core larger and larger would eventually return it to a multimode fiber, but the principal limiting factor is the larger dispersion slope that a larger core area brings. A larger dispersion slope means that the amount of dispersion will be very different according to which wavelength is being used.

Manufacture of more sophisticated single mode fibers requires extensive engineering of the optical preform to give a complex shape to the refractive index profile of the core. Figure 2.27 shows the refractive index profile of large effective area non-zero dispersion shifted fiber, LEA-NZDF.[2] The profile has to be manufactured as accurately as possible to avoid building polarization mode dispersion into the end fiber. Figure 2.28 shows another fiber profile, trapezoid plus ring,[3] which has been optimized for high bit rate, large effective area and low chromatic dispersion at 1550 nm. This particular fiber has a dispersion of 8 ps/nm/km with an effective core area of 65 square microns and has been demonstrated to be able to transmit 150 channels of 10 Gb/s data (i.e. 1.5 Tb/s) in the C and L band with a 50 GHz channel spacing.

Designing long-haul single mode optical systems is now a complex economic exercise that has to balance competing technologies with an acceptance of the reality of the existing installed base of standard 1310 nm single mode fiber.

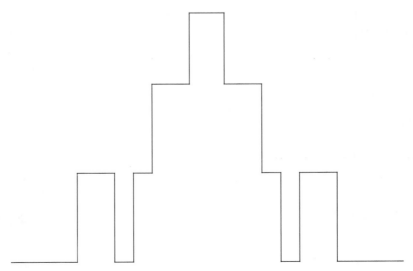

Figure 2.27 *Refractive index profile of LEA-NZDF (preform)*

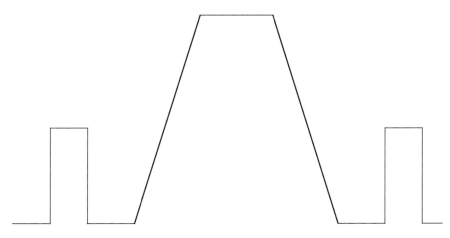

Figure 2.28 *Trapezoid plus ring refractive index profile*

Long-haul submarine optical systems represent possibly the greatest technical challenges and especially demonstrates the technical complexities of fiber choice. For example, one may pick Corning's Submarine SMF-LS™+ which has a negative dispersion −2.0 ps/(nm.km) at 1560 nm to reduce non-linear effects. There is also Corning Submarine LEAF™ fiber with 30% more effective core area and a similar negative dispersion slope at 1560 nm; or there is Corning Submarine SMF-28™+ with a positive 17 ps/(nm.km) slope at 1560 nm which may be used as a dispersion compensating fiber in repeatered systems or a transmission fiber in non-repeatered systems.[4]

Submarine fibers are also strength tested to a higher level, at 200 kpsi (1.379 GPa) compared to the more usual terrestrial standard of 100 kpsi (0.689 GPa).

A brand-new style of optical fiber is known as photonic bandgap or photonic crystal fiber. This new fiber is essentially a length of silica with a honeycomb of hexagonal, or other shape, holes running longitudinally to the axis of the fiber. The mode of propagation is a form of modified total internal reflection, which relies on the geometry of the waveguide, not the refractive index. These short length fibers will have numerous uses, such as tuneable filters, dispersion compensation fibers, short-wave soliton transmission components and polarization control fibers.

For local area networks and premises cabling systems, conventional 1310 nm single mode fiber will remain the most viable and economic single mode solution for some time yet. Specifications are in place for this kind of single mode fiber for use in gigabit Ethernet, IEEE 802.3z, ten gigabit Ethernet, IEEE 802.3ae and ATM and also Fiber Channel.

Plastic optical fiber

Plastic fiber works in the same manner as glass optical fiber but uses plastic instead of glass and usually has a much larger core area. The large core area and easy-to-cut and terminate properties of plastic optical fiber have long held the promise of a low cost, easy to install communications medium that offers all the benefits of optical fiber with the ease of termination of copper. Unfortunately plastic fiber is not yet proven to be cost competitive or to exhibit sufficiently high bandwidth or low enough attenuation to make it a serious rival to either glass fiber or copper cable. Plastic fiber continues to be developed, however, and has found some applications in the automotive field and may yet offer a viable product for short-distance, lower-speed data communications, perhaps in the small office, home office, or SoHo arena.[5]

Plastic fiber available today is step index, which by its very nature limits the bandwidth available. Current designs are based on a material called PMMA, poly methyl methacrylate. Step index plastic optical fiber, or SI-POF, today has a best bandwidth of 12.5 MHz.km and an attenuation of 180 dB/km. Compare this to the 500 MHz.km bandwidth and 1 dB/km attenuation available from 50/125 glass optical fiber.

The manufacturing costs of PMMA fiber are thought to be about the same as for conventional glass optical fiber, but SI-POF currently sells at a premium compared to glass or all-silica fiber. The thermal stability of PMMA is also questionable. High temperatures combined with high humidity can raise the attenuation of the fiber significantly.

SI-POF fibers are available in sizes of 500, 750 and 1000 micron total diameter. Most of this is a PMMA core with a thin layer of fluorinated PMMA for the cladding.

PMMA has attenuation minima occurring at 570 nm and 650 nm. The theoretical minimum attenuation achievable at these wavelengths is 35 and 106 dB/km respectively. 650 nm (red) devices have long been available and now 520 nm (green) LEDs have been produced in the laboratory (NTT and Tohoku University, 2000) enabling a 30 Mb/s signal to be sent over 100 metres of plastic optical fiber.

Deuterated PMMA has been proposed as an advancement. It can reduce attenuation to 20 dB/km in theory but this has not been achieved in practice. Deuterated PMMA is also very expensive to produce.

To really improve plastic fiber a graded index version has to be produced to overcome the poor bandwidth properties of SI-POF. Graded index plastic optical fiber, or GI-POF, offers the potential of 3 Gb/s transmission over 100 m and 16 dB/km attenuation at 650 nm. Even 1300 nm operation may be possible with next generation materials.

GI-POF experiments have been undertaken based on a material called Perflourinated plastic, PF. PF fibers could have an attenuation as low as

1 dB/km at 1300 nm with a fiber of about 750 micron diameter and a 400 micron core. PF is still expensive and nobody has achieved a mass production version of this fiber at such low levels of attenuation, although many laboratory experiments have given encouraging results. Asahi produce a 'PFGI-POF' fiber (CYTOP-Lucina) achieving 50 dB/km and are aiming at 10 dB/km.

Today plastic fibers are mostly used for illumination or very short-distance communication systems, such as in a car. The main advantage of plastic fiber is ease of connectorization but it has yet to prove itself in terms of cost, bandwidth, attenuation and long-term thermal stability.

The ATM Forum has approved a standard for 155 Mb/s over 50 metres of plastic optical fiber. The 980 micron core of POF with 0.5 NA is still seen by many as the easy-connectorization cable media of the future, with 400 Mb/s links over 100 metres being claimed as commercially achievable. The IEEE 1394 Firewire digital bus may well prove to be a viable application for POF, with a speed requirement of 100, 200 and 400 Mb/s over 4.5 metre links. A POF solution could offer 250 Mb/s over 50 metres, more than enough for most equipment interlinks around the average dwelling.

General Motors and Daimler-Chrysler are developing plastic fiber-based automotive audio, video and data distribution systems such as D2B (domestic digital bus) for use in cars, and the next generation of aircraft may well opt for plastic fiber for seat-back video distribution. Any use of optical fiber in an aircraft offers massive weight saving over the equivalent copper cable solution.

References

1   Pondillo P., Lasers in LANs: the effect on multimode-fibre bandwidth, *Lightwave*, November 2000. [*See also TIA Fiber Optic LAN Section www.fols.org*].
2   Tiejun Wang, Mu Zhang, Study on PMD of Large Effective Area G655 Fiber, *Proceedings, International Wire and Cable Symposium* No. 49. Atlantic City, November 2000.
3   Montmorillon, L. et al., Optimized Fiber for Terabit Transmission, *Proceedings, International Wire and Cable Symposium* No. 49. Atlantic City, November 2000.
4   Corning® Product Line Information Sheet PI 1266, *Corning Submarine Optical Fibers*, September 2000.
5   Elliott, B., Blythe, A., Dalgoutte, D., Potential Applications for Plastic Optical Fibre in Datacommunication Networks, *Proceedings, Plastic Optical Fiber Conference POF 97*, Hawaii, September 1997.

3 Optical fiber production techniques

Introduction

There are various methods of manufacturing optical fiber, but all involve the drawing down of an optical fiber preform into a long strand of core and cladding material. This chapter reviews the methods which have been developed to achieve the desired tolerances. As it is drawn down to produce the fiber structure the strand is stronger (in tension) than steel of a similar diameter; however, the silica surface is susceptible to attack by moisture and other airborne contaminants. To prevent degradation the fibers are coated, during manufacture, with a further layer known as the primary coating.

As a result the final product of the fiber production process is primary coated optical fiber. This element is the basis for all other fiber and cable structures as will be seen later.

Manufacturing techniques

The fundamental aim of optical fiber manufacture is to produce a controlled, concentric rod of material comprising the core and cladding. The quality of the product is determined by the dimensional stability of the core, the cladding, and also the position of the core within the cladding.

This chapter reviews the methods by which attempts were made to maximize the control over the above parameters since the early days of fiber production in the late 1960s and early 1970s.

The manufacturing process divides into two quite distinct phases: first the manufacture of the optical fiber preform; second the manufacture of the primary coated optical fiber from that preform.

Irrespective of the quality of fiber manufacture it is impossible to produce high-quality optical fiber from poor-quality preforms. The preform acts as a template for fiber construction and is essentially a solid rod much larger than the fiber itself from which the final product is drawn or pulled much in the same way that toffee can be manipulated from a large block into strands under the action of warm water or air. The preform contains all the basic elements of core and cladding, and the final geometry (in terms of aspect ratio) and numerical aperture are all defined within the preform structure. Thus poor quality control at the preform manufacturing stage results in severe problems in achieving good quality, consistent fiber at the production stage.

In the early days of optical fiber production much attention was given to improving the quality of the preform.

Preform manufacture

Historically there have been three generic methods of manufacturing optical fiber preforms. The first two, rod-in-tube and double crucible, are strictly limited to the production of stepped index, all-glass fibers. As discussed in the previous chapters, large core diameter, high NA fibers were used in the early development of specialist and military systems. Few of these fiber styles find common acceptance in the data or telecommunications fields. A third generic method was investigated and subsequently developed which involves the progressive doping of silica-based materials to create a refractive index profile within the core–cladding formations suitable for the onward manufacture of graded index optical fibers.

Stepped index fiber preforms

Rod-in-tube

The simplest method of producing a core–cladding structure is to take a glass tube of low refractive index, and place it around a rod of higher-index material and apply heat to bond the tube to the rod (Figure 3.1). This creates a preform containing a core of refractive index n_1 surrounded by a cladding of refractive index n_2 which can be subsequently processed.

This method originated in the production of fiber for visible light transmission. To produce fibers capable of transmitting data, high-purity glasses were used, and to produce large core diameter, high NA fibers these methods are still applicable. However, another method suitable for the production of these fiber geometries generated considerable competition for this approach.

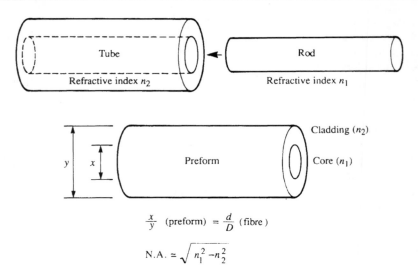

Figure 3.1 Rod-in-tube preform manufacture

Double crucible

One of the main disadvantages of the rod-in-tube process is the necessity for the provision of high-quality glass materials in both the rod and tube formats. The double crucible method overcame this drawback by using glass powders which are subsequently reduced to a molten state prior to their final combination to create the fiber preform.

Figure 3.2 shows a typical arrangement where two concentric crucibles are filled with powdered glasses. The outer crucible contains glass of a low refractive index while the central crucible contains glass having a suitable refractive index to produce the fiber core.

Heat is applied to both crucibles to form melts which are then drawn down to create a fused preform having the desired aspect ratio and numerical aperture.

Both rod-in-tube and double crucible methods were, and are, adequate for the production of large core diameter, high NA optical fibers. However, the processes have obvious limitations.

All-silica fiber preforms

The glass fibers mentioned above with large core diameter, high NA constructions all exhibit relatively high optical attenuation values together with low bandwidths. This is, in part, due to their construction having

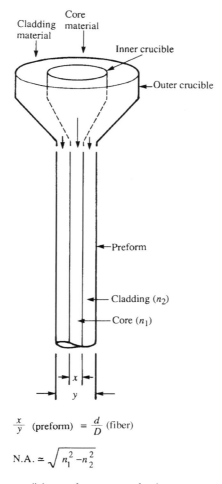

Figure 3.2 Double crucible preform manufacture

been produced in a stepped index configuration with high numerical aperture, but the individual glass materials have relatively high material absorption characteristics.

Silica, better known as quartz, is pure silicon dioxide (SiO_2) and is relatively easy to produce synthetically. It also exhibits low levels of attenuation. It was a natural candidate for the production of optical fibers but has one drawback: it has a very low refractive index, which means that it has to be processed in a special fashion before it is made into a preform.

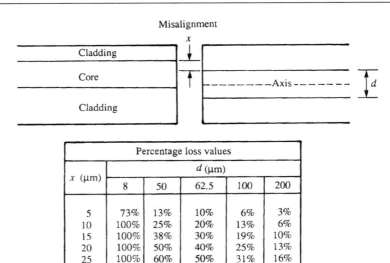

Figure 3.3 The impact of core misalignment

Small-core, graded index fibers

In the previous chapter it was shown that the hunger for bandwidth could be met in two ways. The first, reduction of intermodal dispersion, led to the introduction of graded index fibers while the concept of single mode transmission required small cores which could retain stepped index structure.

Both of these solutions were beyond the capability of the rod-in-tube and double crucible methods of preform production.

Although the optical performance of fibers may not be significantly affected by minute variations in core diameter, the performance of that fiber in a system where it must be jointed, connected, launched into and received from, is very dependent upon the tolerances achieved at the fiber production stage. In Chapters 4 and 5 it is shown that losses at joints and connectors are highly dependent upon the core diameter and its position within the cladding. Mismatches of core diameter and numerical aperture must be carefully assessed before designing any operational highway. A dimensional tolerance of 5 microns which might lead to acceptable mismatches in fibers of core diameter of 200 microns will have a severe impact on a fiber with a core diameter of only 50 microns. Needless to say, such a mismatch would render single mode systems virtually inoperable. Figure 3.3 illustrates this dependence.

To overcome these process limitations new methods had to be developed to manufacture smaller core fibers. Also the need to produce fibers with a varying refractive index across the core dispensed with a two-level construction approach. To provide a solution to these twin requirements the concept of vapour deposition was introduced.

Vapour deposited silica (VDS) fibers

The vast majority of all optical fibers in service throughout the world today have been manufactured by some type of vapour deposition process. There are three primary processes each of which features both advantages and disadvantages.

For simplicity this section begins with a discussion of the inside vapour deposition (IVD).

The ready availability of pure synthetic silica at low cost is key to the VDS production techniques. As already mentioned the refractive index of pure silica is very low, so low in fact that it is difficult to find a stable optical material which has a lower index. This made silica an ideal material for the cladding of fibers but restricted its use as a core compound.

In one version of the IVD process a hollow tube as large as 25 mm diameter forms the basis of the preform. This tube is placed on a preform lathe (see Figure 3.4). The lathe rotates the tube while a burner is allowed to traverse back and forth along the length of the tube. Gases are passed down the tube which are subsequently oxidized on to the inner surface

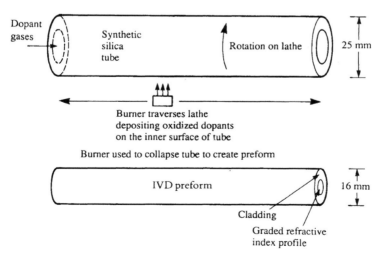

Figure 3.4 *IVD preform manufacture*

of the tube by the action of the burner. The continual motion of the burner and the rotation of the tube enable successive layers of oxidation to be built up in a very controlled fashion. To create the required higher-value refractive index layer (stepped index) or layers (graded index) the composition of the gases is modified with time. IVD is also referred to as MCVD, modified chemical vapour deposition: another version of IVD, termed PCVD (plasma chemical vapour deposition), uses a microwave cavity to heat the gases within the tube. The gas composition is varied by the addition of dopants such as germanium, the inclusion of which increases the refractive index of the deposited silica–germanium layer. Once the desired profile is produced inside the tube the burner temperature is increased and the tube is collapsed to form a rod some 16 mm in diameter. This highly controllable process can create preforms capable of providing finished optical fiber with extremely good dimensional tolerances for fiber core concentricity and excellent consistency for numerical aperture. The main disadvantage of this method is the limited length of the preforms produced which is reflected in the quantity of fiber produced from that preform.

To overcome this limitation other methods have been developed such as outside vapour deposition (OVD) and vapour axial deposition (VAD) which are capable of producing larger preforms. These methods produce preforms of equivalent quality but differ in that a continuous process is adopted and dopants are applied externally rather than internally as discussed above. The preforms produce optical fibers of equivalent optical performance; however, their physical compatibility must be assessed, particularly with regard to fusion splicing as a means of connecting the fibers together. Both the OVD and VAD methods of preform production are more complex than the IVD process.

Outside vapour deposition (OVD)

A graphite, aluminium or silica seed rod is used (Figure 3.5). Using a burner and a variable gas feed, layers of doped silica are deposited on the external surface of the rod. As with IVD, the dopant content is varied to produce the desired refractive index profile; however, in this process the highest dopant (highest refractive index) layers are deposited first, immediately next to the rod. Once the core profile has been produced, a thin layer of cladding material is added, at which point the original seed rod is removed.

The partially completed preform tube is then dehydrated before being returned to the lathe, at which point the final cladding layer is added. A final dehydration process is undertaken to drive out any residual moisture.

The tube is then collapsed down to create the final preform. A variation on this technique involves the use of a prefabricated silica tube, which

Optical fiber production techniques 61

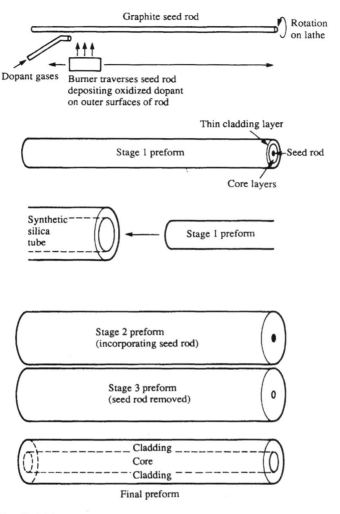

Figure 3.5 *Outside vapour deposition preform manufacture*

is bonded to the partially formed structure to form the cladding layer prior to collapse.

Vapour axial deposition (VAD)

The VAD method utilizes a more complex process but, as will be seen later, offers some advantages.

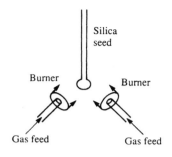
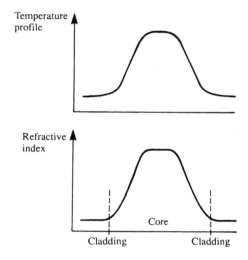

Figure 3.6 *VAD seed, burner configuration and growth mechanism*

Figure 3.6 shows a glass seed together with the other apparatus necessary to produce a VAD preform. It is established fact that the proportion of dopant oxidized on the surface of the seed is dependent upon the temperature of the burner. Using this fact as a foundation, the VAD process achieves the required refractive index profiles by varying the dopant deposition ratio within the core using differences in burner temperature across the seed face. This is shown in diagrammatic form in Figure 3.7. As a result the preform can be produced in a continuous length.

Once the required length of preform is produced then the element is removed from the fixture and dehydrated prior to the application of a cladding tube as discussed above.

Optical fiber production techniques 63

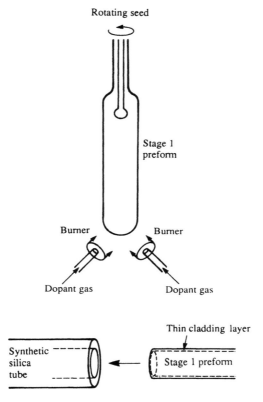

Figure 3.7 *VAD seed growth and preform manufacture*

The IVD, OVD and VAD processes all produce fibers with high performance characteristics; however, each method has its advantages and disadvantages. The principal advantage of VAD is the continuous nature of the preform production, which leads to higher yields at the fiber manufacturing stage. This naturally leads to potentially lower unit fiber costs. Nevertheless IVD and OVD processes tend to offer better core concentricity tolerances due to their integrated structure.

Fiber manufacture from preforms

Figure 3.8 shows a typical arrangement for the drawing of optical fiber from a preform.

The preform is heated in a localized manner and the optical fiber is drawn off or 'pulled' by winding the melt on to a wheel. A fiber

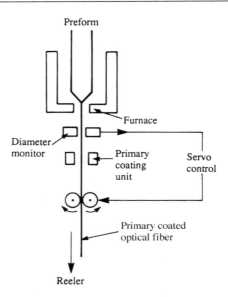

Figure 3.8 *Production of primary coated optical fiber*

diameter measuring system is connected directly by servo-controls to the winding wheel. A tendency to produce fiber with cladding diameter larger than specification is met by an increased drawing rate whilst a small fiber measurement serves to reduce the rate of winding. In this way the cladding diameter is controlled within tight tolerances; the limitations being the drawing rate (as high as possible to produce low-cost fibers) balanced by the response of the servo-control mechanism.

Obviously the highest grade optical fiber is produced from highest grade preforms. Table 3.1 shows the typical physical and optical tolerances achievable for the professional VDS fiber geometries.

Cost dependencies of optical fiber geometries

The cost of optical fiber has always been a key factor in the acceptance of the technology. For that reason it is valuable to understand the dependencies which produce the finished cost of the fiber elements. However, it should be pointed out that in an emerging technology the price of any item may not be directly linked to its cost but rather to any number of market strategies and corporate aims.

The cost of a metre of optical fiber is principally dependent upon:

- preform cost;
- preform yield (length of usable fiber per preform);
- pulling or drawing rate and production efficiency.

Table 3.1 *Optical and mechanical specifications of VDS fibers*

Core diameter microns	Cladding diameter microns	Numerical aperture	Attenuation dB/km 850 nm	Attenuation dB/km 1300 nm	Bandwidth MHz.km 850 nm	Bandwidth MHz.km 1300 nm
8.2	125 ± 1	0.14	—	0.4	—	> 10 000
50 ± 3	125 ± 2	0.2 ± 0.015 2.5		0.8	400	600
62.5 ± 3	125 ± 2	0.275 ± 0.015 3.5		1.5	160	500
100 ± 3	140 ± 3	0.29 ± 0.015 5.0		2.0	100	300

When preforms are produced using the rod-in-tube or double crucible then the costs are based upon the materials used; however, for the widely used VDS fibers the preform cost is dependent upon the numerical aperture of the fiber to be produced. This is because the NA is a measure of the difference in refractive index between the core and cladding structures. The greater the NA, the greater the difference. As the increases in refractive index are produced by expensive dopants and by graded index profiles which take time to produce, then the relationship between the cost of VDS fibers and their numerical aperture is clear.

A further factor which impacts the cost of the preform is its manufacturing yield. Preforms with high dopant content tend to be rather brittle, which reduces the yield of acceptable product.

A preform for the manufacture of 8/125 micron single mode optical fiber (NA = 0.11) is therefore cheaper to construct than a multimode 62.5/125 micron fiber with a numerical aperture of 0.275. At the finished fiber stage the actual ratios are subject to market forces but in general high NA fibers do cost more.

Once manufactured, the preform will be processed into optical fiber and obviously the quantity of end-product has a direct bearing on its cost. Fibers with large cladding diameters are naturally more costly to produce since the amount of usable fiber generated from the preform is correspondingly less.

It is not uncommon to be able to produce 45 000 m of 125 micron cladding fiber from an IVD preform and as much as 400 000 m has been produced from a VAD preform. Equally it is possible to manufacture as little as 600 m of 300 micron cladding fiber from high NA preforms. Naturally the price of such large fibers reflects this fact.

Finally the process efficiency also affects the final cost of optical fiber into the market. A fiber drawing facility is forced to operate continuously and the time spent in changing preforms, manufacturing settings or repairing failed mechanisms is an overhead on the production cost. Larger preforms, reducing change-over times, are an obvious way of reducing the costs by increasing efficiency. This tends to favour OVD or VAD processes.

The price of single mode 8/125 micron fibers is approximately one-third that of multimode products and is likely to fall yet further. The main disadvantage is that the injection of light into the small core necessitates the use of comparatively expensive transmission devices. However, this is unlikely to be insurmountable and efforts are under way to achieve lower cost single mode transmission with devices such as VCSELs. At that time the widespread use of multimode optical fiber geometries will be under threat as new installations adopt the ultimate transmission medium.

Fiber compatibility

As discussed briefly above there are three principal preform production techniques: OVD, IVD and VAD. While producing optical fiber which is optically compatible the different methods result in physically different structures. In particular the viscosity and melting point of the two fibers varies, which makes them more difficult to joint using the fusion splicing method (discussed in Chapters 5 and 6). Although jointing is not impossible, more care has to be exercised and it is therefore important to know the origin of the fiber to be installed, and therefore to select the correct programme on a fusion splicer, in order to avoid unfortunate surprises and failing splices.

Clad silica fibers

When rod-in-tube and double crucible processes were producing relatively stable optical fiber, despite its performance limitations, it was the aim of the industry to develop lower-cost methods.

One development path led to the graded index fibers manufactured by the VDS processes and eventually to the telecommunications grade single mode fiber.

The desire for lower-cost optical fiber to meet early system needs led to the production of plastic clad silica (PCS) fibers. These are produced by taking a pure silica rod preform and drawing it down into a filament (normally having a 200 micron diameter) whilst surrounding it with a plastic cladding material of lower refractive index. It has already been mentioned that it is not easy to find a stable optical material with a lower refractive index than silica and as a result the plastic cladding used was frequently a silicone with a 560 micron diameter. Compared to both the rod-in-tube and double crucible methods the fibers produced were inherently lower cost, but unfortunately the unstable cladding material created its own set of problems which are more fully discussed in Chapter 5.

A much more satisfactory development from this technique has been the production of hard clad silica (HCS) fibers, where the unstable plastic cladding has been replaced with a hard acrylic material which enables the fibers thus produced to be handled in a much more conventional fashion and with acceptable levels of environmental stability.

PCS is now an expensive solution in common with all large core diameter, high NA geometries, and has been replaced by professional grade fibers such as 50/125 micron and 62.5/125 micron. HCS is used in short-distance, data acquisition installations.

Plastic optical fiber

The large core area and easy-to-cut and terminate properties of plastic optical fiber have long held the promise of a low-cost, easy-to-install communications medium that offers all the benefits of optical fiber with the ease of termination of copper. It was once presumed that plastic fiber could also be manufactured at an extremely low price compared to glass or even copper cables. Unfortunately plastic fiber is not yet proven to be cost competitive or to exhibit sufficiently high bandwidth or low enough attenuation to make it a serious rival to either glass fiber or copper cable. Plastic fiber continues to be developed, however, and has found some applications in the automotive field and may yet offer a viable product for short-distance, lower-speed data communications, perhaps in the small office, home office, or SoHo arena.

Plastic fiber available today is step index, which by its very nature limits the bandwidth available. Current designs are based on a material called PMMA, poly methyl methacrylate. Step index plastic optical fiber, or SI-POF, today has a best bandwidth of 12.5 MHz.km and an attenuation of 180 dB/km. Compare this to the 500 MHz.km bandwidth and 1 dB/km attenuation available from 50/125 glass optical fiber.

The manufacturing costs of PMMA fiber are thought to be about the same as for conventional glass optical fiber, but SI-POF currently sells at a premium compared to glass or all-silica fiber. The thermal stability of PMMA is also questionable. High temperatures combined with high humidity can raise the attenuation of the fiber significantly.

SI-POF fibers are available in sizes of 500, 750 and 1000 micron total diameter. Most of this is a PMMA core with a thin layer of fluorinated PMMA for the cladding.

Deuterated PMMA has been proposed as an advancement. It can reduce attenuation to 20 dB/km in theory but this has not been achieved in practice. Deuterated PMMA is also very expensive to produce.

To really improve plastic fiber a graded index version has to be produced to overcome the poor bandwidth properties of SI-POF. Graded index plastic optical fiber, or GI-POF, offers the potential of 3 Gb/s transmission over 100 m and 16 dB/km attenuation at 650 nm. Even 1300 nm operation may be possible with next generation materials.

GI-POF experiments have been undertaken based on a material called perfluorinated plastic, PF. PF fibers could have an attenuation as low as 1 dB/km at 1300 nm with a fiber of about 750 micron diameter and a 400 micron core. Perfluorinated graded index plastic optical fiber, PFGI-POF is available today offering a minimum of 50 dB/km around 1300 nm and 200 dB/km at 650 nm. For example, Lucina® from Asahi Glass, which has a 120 micron core and 500 micron cladding with a numerical aperture of 0.18 and a claimed bandwidth of 200 MHz.km. Manufacturers are aiming for 10 dB/km across the useable spectrum in the near future.

Today plastic fibers are mostly used for illumination or very short-distance communication systems, such as in a car. The main advantage of plastic fiber is ease of connectorization but it has yet to prove itself in terms of cost, bandwidth, attenuation and long-term thermal stability.

Radiation hardness

Radiation hardness is a term indicating the ability of the optical fiber to remain operational under the impact of nuclear and other ionizing radiation. It is not a well-understood subject outside a select band of military project engineers and their design teams. In the commercial market the need for transmission of data under conditions of limited irradiation has led to optical fiber being disregarded despite its suitability for other reasons. Frequently this rejection of the technology has occurred because of lack of valid information; not surprising in an area where 'restricted' information abounds and knowledge is scarce.

As mentioned in Chapter 2 the absorption of light within the core of an optical fiber can be increased under conditions of irradiation. High-energy radiation such as gamma rays can create 'colour centres' which render the fiber opaque at the operating wavelengths of normal transmission systems. Upon removal of the incident radiation the 'colour centres' may disappear and the fiber may return to its original condition and its performance may be unaffected.

It will be noticed that the above paragraph contains many 'cans' and 'mays'. This is because all fibers react differently to irradiation and as a result 'radiation hardness' with regard to optical fiber remains a vague term. This section serves to explain the issues of radiation hardened optical fiber and aims to define a pathway through the jargon.

A particular fiber can be hardened against specific levels of radiation but the entire system, including the transmission equipment, must be assessed in terms of its true 'hardness' requirements. Obviously the transmission will be disrupted when the network attenuation exceeds a specified limit. This represents an increase over the normal attenuation of the cabled fiber which may be attributed to the impact of radiation incident on the fiber core. In this way a radiation performance requirement can be generated:

If Dose A is incident upon the fiber for a period B then the resulting attenuation shall not increase by more than C.

Some users may accept system failure during irradiation as long as there is a known recovery state, and this will mean another specification for the recovery time, e.g.

If Dose A is incident upon the fiber for a period B and subsequently removed, then after time C the resulting attenuation shall not be more than D.

From the above radiation performance requirements it would appear that there is no such thing as 'radiation hardness' as an all-encompassing parameter and as a result a fiber is neither 'radiation hard' nor 'radiation non-hard' but has a specified performance against given levels of radiation.

The optical fibers manufactured from the range of materials and preform techniques discussed above offer a variety of performance levels under the influence of radiation. The radiation performance requirement generated for a given system can therefore be matched against the known performance of these fiber designs.

Pure silica fibers such as PCS and HCS designs exhibit moderate radiation hardness. It is the addition to pure silica of dopants such as germanium that encourages the formation of 'colour centres' which are responsible for the attenuation increases observed.

This suggests that the higher bandwidth, lower attenuation, fiber geometries manufactured from VDS preforms exhibit lower levels of 'radiation hardness' – a feature which is observed in practice. To overcome this problem, further preform dopants are added during the deposition process. The dopants, such as boron and phosphorus, act as buffers preventing the formation, or speeding the removal, of 'colour centres'. In this way a range of VDS-based fibers have been rendered radiation hard against a particular radiation performance requirement.

As the germanium content is directly linked to the numerical aperture of the fiber it is interesting to note that radiation performance improves for lower NA values. It is realistic to expect 8/125 micron single mode fibers to exhibit considerably better radiation performance than their multimode counterparts. This is observed in practice.

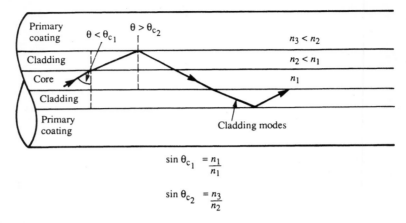

Figure 3.9 *Cladding mode transmission*

Primary coating processes

In Figure 3.8 the drawing of optical fiber is shown accompanied by a secondary process — the addition of a primary coating which immediately surrounds the cladding and prevents surface degradation due to moisture and other pollutants.

When produced, the optical fiber is stronger than steel in tension but unfortunately humidity can rapidly produce surface defects and cracks on the silica cladding which can eventually lead to complete failure by crack generation if not controlled.

The primary coating is normally a thin, ultraviolet cured acrylate layer with a diameter of around 250 microns, depending upon the fiber geometry. The layer is the subject of particular interest because it immediately surrounds the cladding. Its refractive index is of concern due to its ability to trap light between the CCI and the cladding surface. At one time primary coatings were applied which had lower refractive indices than the cladding. Light therefore becomes trapped by TIR as is shown in Figure 3.9. These cladding modes, whilst not impacting overall attenuation, do cause confusion at the test and measurement stage.

Generally these primary coatings have been phased out and they have been replaced by mode stripping fibers with high index coatings. These absorb the cladding modes rapidly.

The primary coating on modern fiber is an easy-to-strip substance yet providing an essential part of the mechanical integrity and water resistance of the optical fiber structure as a whole.

Summary

The range of optical fiber geometries necessary to meet both current and historic demands can be serviced by a variety of manufacturing techniques.

Large core diameter, high NA fibers can be produced from pure silica (HCS), doped silica (VDS) or glass components (double crucible).

The lower NA, smaller structured fibers required tight tolerance preforms and the VDS processes meet this need.

The cost base is well defined with large core diameter, high NA fibers costing significantly more than the high-performance single mode products.

The end result of all fiber production is primary coated optical fiber which is the foundation for all further cabling processes.

4 Optical fiber – connection theory and basic techniques

Introduction

Having covered optical fiber theory and production techniques in the previous two chapters it may seem natural to move to optical fiber cable as the topic of this chapter. However, before leaving optical fiber it is relevant to cover the connection techniques used to joint optical fiber in temporary, semi-permanent or permanent fashions. The theoretical basis of fiber matching and, more importantly, mismatching, is of vital importance in many areas of cabling design and demands treatment ahead of the purely practical aspects of cabling.

Connection techniques

The end result of the fiber production process is primary coated optical fiber (PCOF). The PCOF is not manufactured in infinite lengths and therefore must be jointed together to produce long-haul systems. For short-haul data communications the PCOF, in its cabled form, may be either jointed or connected at numerous points. Equally importantly the cabling may have to be repaired once installed and the repair may require further joints or connections to be made.

Therefore to achieve flexibility of installation, operation and repair it is necessary to consider the techniques of connection as they apply to optical fiber.

As will be seen the connection techniques are inevitably linked to PCOF tolerances and acceptable performance of joints and interconnection is the result of careful design and not pure chance.

The major difference between copper connections and optical fiber joints is that a physical contact between the two cables is not sufficient. The passage of current through a 13 amp mains plug relies purely on good physical (electrical) contact between the wires and the pins of the plug. To achieve satisfactory performance through an optical fiber joint it is necessary to maximize the light throughput from one fiber (input) to the other (output).

Optical fiber connection techniques are frequently of paramount importance in cabling design because, perhaps surprisingly, the amount of transmitted power lost through a joint can be equivalent to many hundreds of metres of fiber optic cable and is a major contributor to overall attenuation. It is therefore important to gain a complete understanding of the mechanisms involved in the connection process and their measurement.

Connection categories

There are many ways to categorize the range of connection techniques applicable to optical fiber. The divisions are a little arbitrary but in this book the two major options are:

- fusion splice jointing; and
- mechanical alignment.

The former is a well-proven technique wherein the fibers are prepared, brought together and welded to form a continuous element which is, in the perfect world, both invisible to the naked eye and to any subsequent optical measurement.

Mechanical alignment on the other hand is a very wide-ranging term covering:

- mechanical splice joint;
- butt joint (non-contacting) demountable connector;
- butt joint (contacting) demountable connector;
- any other technique not covered above.

This chapter concentrates upon the loss mechanisms encountered in jointing optical fibers whilst Chapter 6 reviews optic fiber connector designs and takes a close look at their assessment against their manufacturers' specifications. It concludes that, in many cases, the connectors currently available are as good as the optical fiber allows them to be. Chapter 6 discusses installation techniques and the suitability of a particular jointing technique to specific installation environments.

Insertion loss

The optical performance of any joint can be measured from two viewpoints – its performance in transmission (that is the proportion of power transmitted from the launch fiber core into the receive fiber core) and its performance in reflection, known as return loss, being the proportion of power reflected from the joint back into the launch fiber core. Insertion loss is the term given to the reduction in transmitted power created by the joint. Both return loss and insertion loss are measured in decibels. Decibels are defined in equation (4.1):

$$\text{insertion loss (dB)} = -10 \log_{10} \frac{\text{power transmitted}}{\text{power incident}} \qquad (4.1)$$

The ideal joint transmits 100% of the launched power and has a 0 dB insertion loss. Most joints fail to achieve this standard and have a positive insertion loss (see Table 4.1).

Table 4.1 *Transmitted power and insertion loss*

Transmitted power (%)	Insertion loss (dB)
100	0
90	0.46
80	0.97
70	1.55
60	2.22
50	3.01

Basic parametric mismatch

Looking for the ideal connection technique the designer is aiming for a zero power loss in the transmitted signal (which implies zero reflected power also). This means that all the light emitted from the core of the first fiber is both received and accepted by the core of the second. This suggests perfect alignment of the two mated optical cores.

Before assessing the performance of real jointing techniques it is worth while to investigate the losses which might be seen through the use of a perfect joint.

As most joint technology uses the cladding diameter as a reference surface it is possible to define the perfect joint as one in which two fibers of equal cladding diameter are aligned in a V-groove or equivalent mechanism (see Figure 4.1). This section looks at the losses generated at this joint by the basic mismatches in optical fiber parameters due to tolerances which result during the manufacture of both the preform and the fiber itself.

Optical fiber — connection theory and basic techniques 75

Figure 4.2 looks at the joint with regard to core diameters d_1 and d_2. The power output P_{out} from a perfectly aligned joint is defined in equations (4.2) and (4.2):

$$P_{out} = P_{in} \quad \text{for} \quad d_2 > d_1$$

$$P_{out} = P_{in} \frac{(d_2)^2}{(d_1)^2} \quad \text{for} \quad d_2 < d_1 \tag{4.2}$$

$$\text{insertion loss} = 20 \log_{10} (d_2/d_1) \tag{4.3}$$

This merely continues the idea of light acceptance being a function of core cross-sectional area.

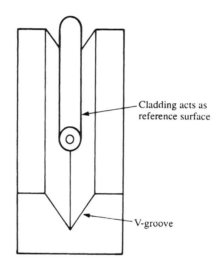

Figure 4.1 *V-groove alignment*

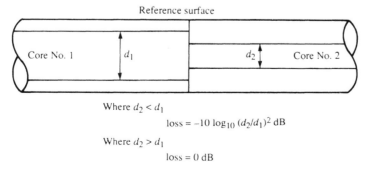

Figure 4.2 *Core diameter mismatch*

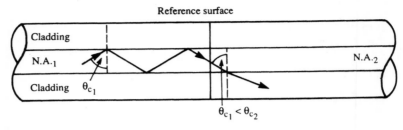

Where $N.A._1 > N.A._2$
$$\text{loss} = -10 \log (N.A._2/N.A._1)^2 \text{ dB}$$

Where $N.A._2 > N.A._1$
$$\text{loss} = 0 \text{ dB}$$

Figure 4.3 *Numerical aperture mismatch*

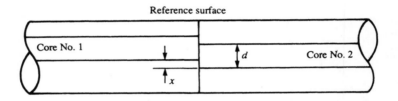

$$\text{Loss} = -10 \log_{10} \left\{ \frac{1}{90} \tan^{-1}\left(\frac{de}{x}\right) - \frac{2xe}{\pi d} \right\} \text{ dB}$$

$$\text{where } e = \left(1 - \frac{x^2}{d^2}\right)^{0.5}$$

Figure 4.4 *Eccentricity mismatch*

Figure 4.3 looks at the joints with regard to numerical aperture and the equations, (4.4) and (4.5), are similar:

$$P_{out} = P_{in} \quad \text{for} \quad NA_2 > NA_1$$
$$P_{out} = P_{in} \frac{(NA_2)^2}{(NA_1)^2} \quad \text{for} \quad NA_2 < NA_1 \quad (4.4)$$

insertion loss $= 20 \log_{10} (d_2/d_1)$

where $NA_1 > NA_2$ then loss $= -10 \log (NA_2/NA_1)^2$ \qquad (4.5)

where $NA_2 > NA_1$ then loss $= 0$ dB

Finally Figure 4.4 assesses the impact of core misalignment due to, perhaps, core eccentricity within the cladding and again based upon acceptance of light being related to cross-sectional area:

$$P_{out} = P_{in} \left(\frac{1}{90} \tan^{-1}\left(\frac{de}{x}\right) - \frac{2xe}{\pi d} \right) \text{ dB} \tag{4.6}$$

Where d = diameter, x = misalignment and $e = (1 - x^2/d^2)^{0.5}$

A graphical representation of this complex equation is shown in Figure 4.5.

These three equations (4.3, 4.5 and 4.6) apply to any joint and again looking at Figure 4.1 it is clear that a perfect joint made with two nominally identical fibers can create significant losses.

The example of 50/125 micron 0.20 NA fiber is examined in detail and Table 3.1 shows the parameter tolerances for the standard optical fiber in terms of the core diameter, its concentricity, the cladding diameter and the numerical aperture.

Core diameter 50 ± 3 microns
d_1 = 53 microns d_2 = 47 microns
Insertion loss = 1.02 dB
Numerical aperture 0.20 ± 0.015
NA_1 = 0.215 NA_2 = 0.185
Insertion loss = 1.31 dB
Core concentricity: 0 ± 2 microns
Insertion loss = 0.47 dB

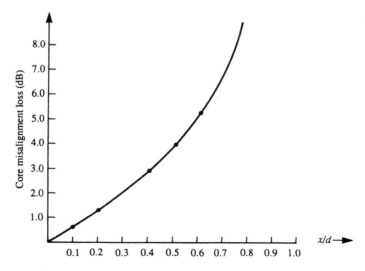

Figure 4.5 *Core misalignment losses*

In total a perfect joint between two identical geometry fibers both within specification can create a combined loss of 2.79 dB.

If the actual losses of otherwise perfect joints were as poor as predicted using the above worst case assumptions, then it would be almost impossible to construct any fiber highway which required patching or repair and a more realistic, statistical approach is investigated in Chapter 5 which suggests a worst case parametric mismatch of 0.31 dB. However, the purpose of pursuing this approach is to furnish the reader with a basic understanding of the losses which can be generated by the parametric mismatches within batches of optical fiber, independent of the joint itself.

Fusion splice joints

The purpose of a fusion splice joint is to literally weld two prepared fiber ends together, thereby creating a permanent (non-demountable) joint featuring the minimum possible optical attenuation (and no reflection). Figure 4.6 illustrates the technique.

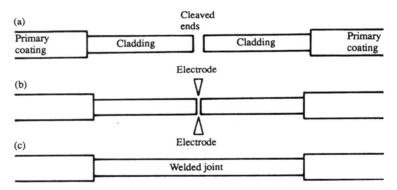

Figure 4.6 *Fusion splice jointing*

The loss mechanisms in such a joint may be summarized as follows:

- *Core misalignment.* Although normally aligned using the cladding diameter as the reference surface, it is generally believed that the complex surface tension and viscosity structures within the core and the cladding do tend to minimize the actual core misalignment.
- *Core diameter.* As previously discussed the allowable diameter tolerances creates the possibility of attenuation within the joint.

    However, where differences between core diameters are large the welding of core to cladding inevitably takes place, which can either exaggerate or reduce the resultant losses. As a logical extension to this

it should be obvious that optical fibers having different geometries are difficult if not impossible to joint using the fusion splice method.
- *Numerical aperture.* The above comments regarding core diameter apply also to numerical aperture mismatches.

In addition to these losses, which are almost unavoidable, the level of skill involved in the process demands few abilities other than those necessary to prepare the fiber ends by cleaving. Nevertheless, poor levels of cleanliness and unacceptable cleaving of the ends will incur additional losses due to the inclusion of air bubbles or cracks. Incorrect equipment settings will also influence losses achieved and may result in incomplete fusion.

Fusion splicing is capable of producing the lowest loss joints within any optical fiber system, but their permanence limits their application. The need for demountable connections necessary to facilitate patching, repair and connection to terminal equipment forces the use of jointing techniques which use mechanical alignment.

Mechanical alignment

The fusion splice outperforms other mechanisms because it integrates the two optical fibers, creating optimum alignment of the optical cores. Any mechanical alignment technique, chosen for reasons of system flexibility, will incur losses in addition to those already discussed for the fusion splice.

The non-fusion splice methods are varied and their adoption depends largely upon the application and the connection environment.

Butt joint (contacting or non-contacting) demountable connectors are normally effected by applying plug (male) terminations to each fiber and subsequently aligning these components within a barrel fitting. The latter are variously described as uniters, adaptors and, rather confusingly, couplers. The term used in this book will be restricted to adaptor only.

The butt joint is most frequently seen on transmission equipment and at patch panels where flexibility is a key requirement.

Mechanical splicing is an alternative to fusion splicing and uses a simple but high-quality alignment mechanism which enables the positioning and subsequent fixing of the two fiber ends by the use of crimps or glue. Frequently the techniques involve some element of light loss optimization.

Inevitably any core diameter or numerical aperture mismatch across the joints will create some degree of attenuation but mechanical alignment techniques may incur further losses due to:

- lateral misalignment: see Figure 4.7;
- angular misalignment: see Figure 4.8;
- end-face separation: see Figure 4.9;
- Fresnel reflection: see Figures 4.10, 4.11 and Chapter 2;

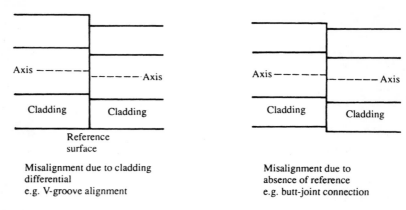

Figure 4.7 Mechanisms for lateral misalignment

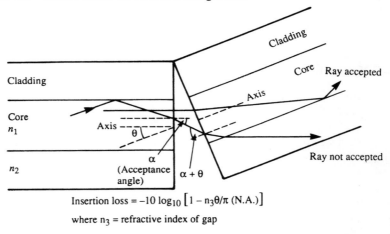

Figure 4.8 Angular misalignment

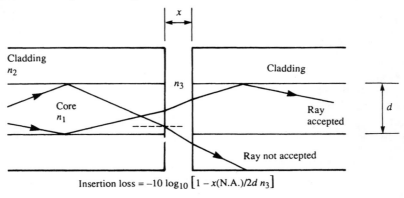

Figure 4.9 End-face separation

- cleave angle not perpendicular to the fiber axis;
- broken or chipped fiber end face;
- dirt on the fiber face.

As mentioned above, it is normal to use the cladding diameter as the reference surface which has a specified tolerance. Any mechanical system of alignment that relies upon the cladding diameter has an inherently greater capacity for misalignment than the fusion joint in which the cladding misalignment tends to be limited due to the surface tension and viscosity effects discussed above. Figure 4.12 emphasizes this point. Lateral misalignment is governed by the same formula as used for the calculation of losses due to core eccentricity.

Another factor in the total misalignment is the effect of angle. Angular misalignment is rarely seen in fusion splice joints but can be significant in certain types of mechanical joint. Figure 4.13 illustrates the effect which is

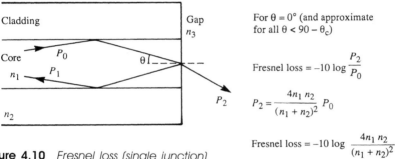

Figure 4.10 *Fresnel loss (single junction)*

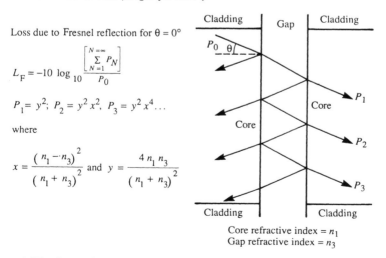

Figure 4.11 *Fresnel loss at a joint*

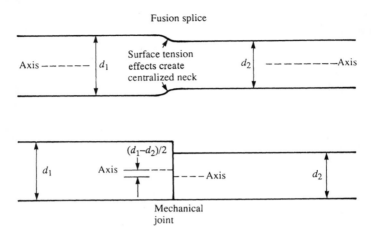

Figure 4.12 *Misalignment: mechanical versus fusion*

most common in older types of butt joint connector. This is not to be confused with the special angled-face connectors designed to prevent reflections (these are discussed at the end of this chapter in the section entitled Return loss). These losses are reduced for high NA fiber geometries, with the insertion loss caused by angular misalignment given in equation (4.7):

$$\text{insertion loss} = -10 \log_{10}\left[\frac{1 - n_3\theta}{\pi NA}\right] \quad (4.7)$$

If a joint is produced with a small gap between the two fiber ends then two further effects will come into play:

- separation spreading;
- Fresnel reflection.

Separation spreading is due to the gap enabling light to spread, within the confines of the acceptance angle cone, as it emerges from the launch

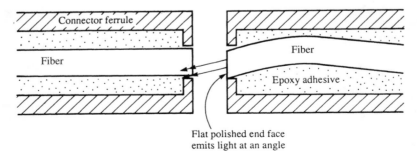

Figure 4.13 *Angular misalignment within connectors*

fiber core and not being captured by the receiving fiber core. This loss is exaggerated for high NA fibers, with end-face separation given in equation (4.8):

$$\text{insertion loss} = -10 \log_{10} \left[\frac{1 - x\text{NA}}{2dn_3} \right] \quad (4.8)$$

Fresnel reflection, on the other hand, is a physical phenomenon as discussed in Chapter 2. If a gap exists between the two optical fibers the Fresnel reflection will occur twice. First as the light leaves the launch fiber core (silica to air) and second as the light is accepted into the receive fiber core (air to silica).

As shown in Figure 4.11, Fresnel reflection reduces the forward transmitted power by a known percentage based upon the equations shown in Figure 2.4 (for $\theta = 0°$):

$$P_{accepted} = P_{incident} \left\{ y^2 \sum_{n=0}^{n=\infty} x^{2n} \right\} \quad (4.9)$$

where $x = \dfrac{n_1 - n_3}{(n_1 + n_3)^2}$ and $y = \dfrac{4n_1 n_3}{(n_1 + n_3)^2}$

For air gaps $n_3 = 1.00027$ and for a typical silica core $n_1 = 1.48$. Using these figures gives:

$$\text{insertion loss (dB)} = -10 \log_{10} \frac{P_{accepted}}{P_{incident}}$$

$$= -10 \log_{10} (0.925 + 0.001 + \ldots)$$

$$= 0.33 \text{ dB}$$

This level of loss is unavoidable where an air gap exists between the fiber ends and, historically, fully demountable connectors could never achieve losses better than this. However, there are two ways in which these losses can be reduced:

- match the refractive index of the gap to that of the fiber core;
- reduce the end-face separation (ideally to zero).

The use of index matching gels is acceptable for semi-permanent joints; however, fully demountable connectors do not favour their use (the gels or fluids become contaminated or dry out, thereby increasing the problems rather than solving them).

More recently there has been a trend towards connectors which feature physical contact between the fiber ends, primarily as a means of reducing reflections. These types of connectors exhibit little or no Fresnel reflection and are generally able to produce significantly reduced overall insertion loss figures. These connectors are more fully discussed later in this book.

Joint loss, fiber geometry and preparation

Table 4.2 and Figure 4.14 indicate the typical losses exhibited by the various joint mechanisms. It is clear that fusion splice joints are able to produce the lowest insertion loss values (both initially and, experience has shown, over a considerable period of time) whereas the losses associated with demountable connectors are significantly greater due to the greater levels of core misalignment, end-face separation etc.

It should be clear that for a given degree of misalignment the larger core, higher numerical aperture fiber geometries offer advantages in terms of the insertion loss resulting for the reason that lateral misalignment of a given value will have less effect as the core diameter is increased.

Table 4.2 *Typical loss values for common connection methods using 50/125 micron, 0.2 NA*

Loss mechanism	Loss (dB)		
	Fusion splice	Mechanical splice	Demountable connector
Parametric	0.31	0.31	0.31
Fresnel loss	—	0.10	0.34
Lateral misalignment	0.19	0.23	0.60
End-face separation	—	0.10	0.25
Angular misalignment	—	0.06	0.20
Total	0.5	0.8	1.7

In all joints, preparation is of paramount importance. Specifically the ability to 'cleave' (cut the end of the fiber perpendicular to its axis) is vital, both in fusion splice and mechanical joints.

Return loss

The power reflected from a joint can be as important as the power transmitted. It is normally termed return loss, is measured in decibels and is defined, with reference to Figure 4.15, in equation (4.10):

$$\text{return loss (dB)} = -10 \log_{10} \frac{P_{\text{reflected}}}{P_{\text{incident}}} \qquad (4.10)$$

The multiple reflections taking place in the air gap between the two fiber end-faces, as shown in Figure 4.16, can be treated as shown in Figure 2.4.

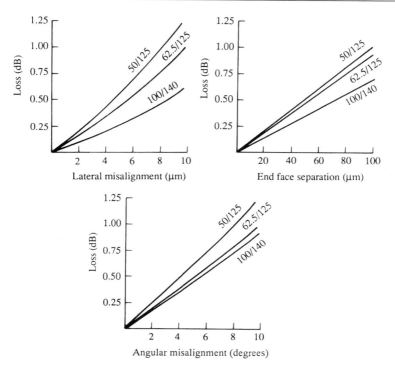

Figure 4.14 *Loss values versus fiber geometry*

The total reflected power is therefore calculated (for $\theta = 0°$) from equation (4.11):

$$P_{reflected} = P_{incident} \left\{ x + xy^2 \sum_{n=0}^{n=\infty} x^{2n} \right\} \qquad (4.11)$$

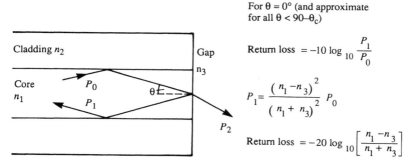

Figure 4.15 *Return loss (single junction)*

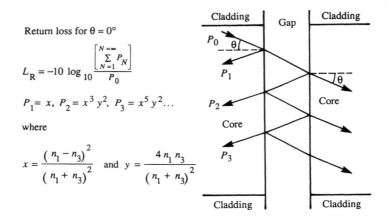

Figure 4.16 Return loss at a joint

$$\text{where} \quad x = \frac{n_1 - n_3}{(n_1 + n_3)^2} \quad \text{and} \quad y = \frac{4n_1 n_3}{(n_1 + n_3)^2}$$

The summation represents the continuing reflections taking place between the end-faces. Calculating in a rough fashion, using the same values as in the Fresnel loss calculation above, gives:

$$P_{\text{reflected}} = P_{\text{incident}}(0.0375 + 0.0347 + 0.0000488 + \ldots)$$
$$= P_{\text{incident}}(0.0722)$$

and return loss = 11.41 dB

Therefore any joint with an air gap is predicted to exhibit a return loss of approximately 11 dB.

Return loss was not a particularly important feature in optical connection theory until high-speed communications using lasers was developed. Lasers do not perform well when subject to high levels of reflected light and performance suffers both in the short and long term. Methods had to be found to reduce the reflections from nearby joints by improving the return loss figures.

As discussed earlier, physical contact between end-faces is now commonplace, with the result that little or no air gap remains. Return loss figures of more than 30 dB are produced. These developments have resulted in improvements in insertion loss by the removal of Fresnel loss

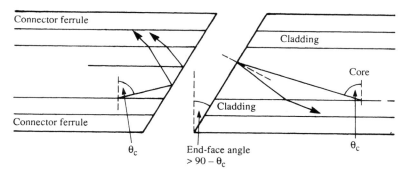

Figure 4.17 *Angled face connectors*

in the forward direction. Unfortunately the contact between the end-faces carries with it an added responsibility for cleanliness and long-term performance can suffer if rigorous instructions are not followed.

Another method of improving the return loss which is receiving warranted attention amongst single mode connector applications is the use of demountable connectors which feature angled ferrule end-faces. These are designed such that the reflections lie beyond the critical angle at the CCI and are removed from the core, as shown in Figure 4.17. These connectors are very effective and in certain applications are the ideal solution.

Summary

This chapter has reviewed connection theory and the limitations underlying any kind of joint.

Insertion loss and return loss are the measures of the optical performance of an optical fiber joint. Their dependence upon core diameter, numerical aperture and misalignment has been discussed and will be further expanded upon in the next chapter.

The methods for the connection of optical fiber vary and all are designed to minimize the wastage of light, but the ideal joint with zero power loss is rarely attainable, since fiber tolerances form the limit rather than the joint mechanism itself.

This idea may not be in line with the product literature generated by the manufacturers of the joint components. They naturally attempt to promote their product in a competitive market by offering premium performance. Written specifications must therefore be carefully studied before accepting their validity in the real world of cabling designs and installations. This is covered in Chapter 5.

5 Practical aspects of connection technology

Introduction

The theoretical analysis of a fiber optic connection with due regard to basic parameter mismatch, misalignment and the other factors discussed in Chapter 4 is useful to understand the losses within the various types of optical fiber joint.

However, at the practical level this theory is submerged in a sea of marketing, standardization and specification jargon. Finally, a goodly amount of processing is involved in producing any of the joints discussed and the quality of the processing may further muddy the waters which, it is hoped, showed moderate clarity at the end of Chapter 4. This chapter seeks to mark out the true path through this most difficult area in an effort to enable the reader to determine how joints *will* perform rather than how they *can* perform.

Alignment techniques within joints

The previous chapter defined two types of joint mechanisms:
- fusion splice techniques;
- mechanical alignment techniques;

For the purpose of the next section it is valid to recategorize all joints as using either:
- relative diameter cladding alignment, or
- absolute cladding alignment.

Relative cladding diameter alignment refers to those techniques which do not depend upon the absolute value of the reference surface (the cladding diameter) but rather are based upon the relative values of the diameter.

Practical aspects of connection technology 89

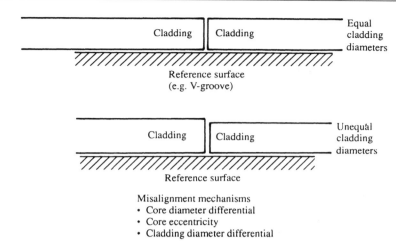

Figure 5.1 *Relative cladding diameter*

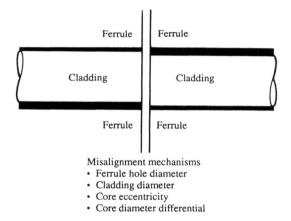

Figure 5.2 *Absolute cladding diameter alignment*

Examples are fusion splice jointing and V-groove mechanical splicing, where provided that the two cladding diameters are equal (independent of their values) then the core alignment will be purely a function of its own eccentricity with the cladding (see Figure 5.1).

Absolute cladding diameter alignment processes involve the use of ferrules as in some mechanical splices and virtually all demountable connectors. The ferrules have holes along the axis to allow the fibers to be brought into alignment. The diameter of these holes and the position of the fiber within them is one further misalignment factor in the complex issue of joint loss and its measurement (see Figure 5.2).

The joint and its specification

The optical specification of any jointing technique is based on its insertion loss and its return loss. As discussed in the previous chapter, the insertion loss of a joint is a measure of the core–core matching and alignment within the joint.

From Figure 5.3 we have equation (5.1):

$$\text{insertion loss} = -10 \log_{10} \frac{P_{\text{accepted}}}{P_{\text{incident}}} \text{ dB} \qquad (5.1)$$

It was continually highlighted in Chapter 4 that any forecast of the insertion loss should take into account basic parametric mismatches of the optical fiber as well as the performance of the connection technique.

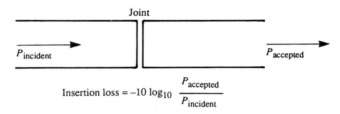

Figure 5.3 *Insertion loss*

The justification for this stance is simple: in real fiber optic cabling it is normal for a single optical fiber geometry to be adopted. Nevertheless any individual link may comprise several types of optical fiber cable – perhaps a direct burial cable jointed to an intra-building cable jointed to an office cable connected via a jumper cable assembly to the transmission equipment (see Figure 5.4). The fiber in each cable will almost always have different origins and batch history despite having the same nominal fiber geometry. All these fibers have to be jointed and therefore it is vital to understand what a joint will produce (in terms of insertion loss) rather than what it can produce (based upon the product literature). The performance that will be produced will depend upon the quality and dimensional tolerance of the two fibers involved as much, if not more, than the submicron accuracy of the connection technique.

Scepticism is therefore the watchword and careful assessment is the prerequisite for understanding the difference between the figures for insertion loss quoted by the joint manufacturers and the potential results obtained with a real system.

Practical aspects of connection technology

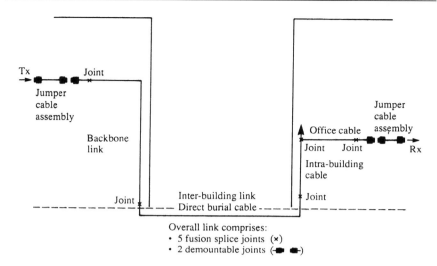

Figure 5.4 *Complex transmission system*

Insertion loss and component specifications

To obtain a fair assessment of 'will' rather than 'can' performance it is necessary to undertake some independent testing of joints on a sample basis.

If a large number of fiber ends are produced using a given fiber geometry from a variety of manufacturing batches and subsequently jointed then a histogram can be produced (Figure 5.5) which will normally appear significantly worse than the product data produced by the manufacturer of the jointing components themselves. The reason for this is that this group of results makes allowance for basic parameter mismatches in the fibers used.

The joint manufacturer's datasheets quote all manner of measurements, but the majority attempt to present an optimistic picture in which truth suffers at the expense of marketing edge in what is certainly a very competitive market place. That being said it is rare for the datasheets to contain untruth and it is left to the unsuspecting cabling designer or installer to discover the validity, or not, of the claims made.

The unacceptable face of joint 'specmanship' can be presented in a number of ways. Some of these are detailed below with the aim of sensitizing the readers and, it is hoped, enabling them to make their own assessment of the published data.

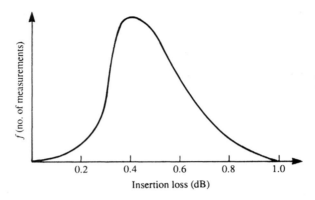

Figure 5.5 *Random mated insertion loss histogram*

The ideal fiber model

The ideal fiber model is a frequently practised technique in which the measurement of joint loss removes any misalignment due to fiber tolerances. There is no allowance for parametric mismatch and the fiber acts as if it were perfectly formed or ideal.

The ideal 50/125 μm fiber behaves as if it had a core diameter of 50 μm (with zero tolerance) and a numerical aperture of 0.20 (with zero tolerance) and a core aligned precisely along the axis of the fiber.

If such a fiber were able to be manufactured in large quantities the joints produced would exhibit considerably better optical performance than is seen in practice due to the absence of basic parametric mismatch.

The manufacturers justify the use of this model by arguing that their responsibility lies in the production of jointing components and not the manufacture of optical fiber. They also suggest that the provision of fiber and the performance of the final joint is the responsibility of the installer. There is some merit in this argument in the sense that the ultimate joint mechanism would exhibit zero insertion loss for the ideal fiber and that the model does offer some measure of joint mechanism capability. However, most manufacturers fail to inform their customers that a realistic performance assessment must include basic parametric mismatches, and therefore tend to undermine the credibility of the other information provided.

Ideal-fiber model measurements

Joints using the ideal-fiber model are measured by the following method. The fiber ends to be jointed originate at the same point within a piece

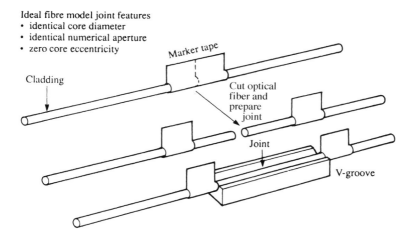

Figure 5.6 *Ideal fiber model: jointing technique*

of fiber of the desired geometry. The fiber is marked with an orientation baseline prior to cutting. After cutting the fiber ends are prepared such that as little fiber as possible is wasted. The joint is then made ensuring that the fibers are connected in their original orientation.

In this way the fibers which have been jointed behave as having identical core dimensions and NA, and the maintenance of orientation acts to minimize any possible core eccentricity (see Figure 5.6).

This method of measurement produces levels of loss which are considerably better than those found if the two fibers to be jointed are selected from two different locations on the same drum or, perhaps more realistically, from different fibers manufactured at different times by different companies.

The interference-fit model

The above method applies most satisfactorily to the fusion splice and V-groove mechanical splice types of joint where the alignment techniques depend upon the equality of cladding diameters rather than their absolute values (see Figure 5.7).

The majority of joint mechanisms do, however, rely upon alignment techniques which are dependent upon absolute values of cladding diameter, e.g. ferrules within demountable connectors. For these types of joints the interference-fit model can be used to significantly distort the true performance figures achieved by eliminating the misalignment due to the effect of cladding diameter tolerances combined with ferrule hole size tolerances.

94 Fiber Optic Cabling

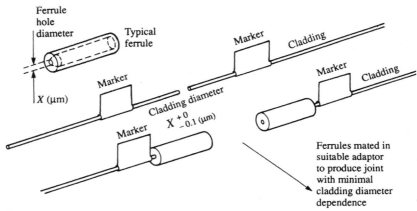

Figure 5.7 Interference-fit model: jointing technique

Interference-fit measurements

This technique involves the measurement of cladding diameter over a considerable length of fiber to locate a position where the diameter will be a close or interference fit within the ferrule hole. The fiber is then treated as an ideal fiber as detailed above.

The misleading results for insertion loss produced by this technique are founded upon the unnatural nature of the experimental set-up. The implications of the misinformation produced by this method go deeper than just the insertion loss. One parameter which is frequently overlooked in the examination of joint specifications is that of rotational variation (the change in insertion loss as one joint face is rotated against the other). The interference fit methods of measurement restrict the degree of misalignment to that related to core eccentricity which is considerably lower than would be found by considering minimum cladding diameter fibers in maximum diameter ferrule holes.

In a multimode connector (made to suit either 50/125 μm or 62.5/125 μm fiber) the ferrule hole may lie between 128 and 129 microns whereas the fiber cladding diameter may lie in the range 122–128 μm. The resultant misalignment could therefore be as much as 7 μm, which is much greater than the maximum fiber core eccentricity of ±2 μm. Any technique of preparation and measurement which ignores this level of misalignment is dubious and cannot be readily accepted in the practical world of installation.

Sized-ferrule techniques

A further complication is produced by certain manufacturers who offer ferrules with a range of hole sizes.

This is common in telecommunications, where fibers are normally sized prior to the selection of ferrules with the appropriate hole diameter. Other techniques for minimizing losses are covered later in this chapter.

However, certain companies seek to offer multiple hole sizes for multimode, data communications, applications. Normally these are 126 μm and 128 μm. Undoubtedly terminated cables which are fitted with 126 μm ferrules will perform better than those with 128 μm alternatives since the misalignment will be correspondingly less. Unfortunately the use of 126 μm ferrules will limit the yield achieved on larger cladding diameter fibers and attempts to recreate the losses achieved on 126 μm ferrules across a broad range of fibers can be an expensive and unhelpful mistake.

The three techniques listed above are tried and tested methods by which results for insertion loss of joints of all types can be improved by rendering them unrealistic. It is therefore important to be able to understand the basis of the figures quoted and even more important it is necessary to understand the likely results under random connection conditions which must include the basic parametric mismatches of the optical fiber itself.

The various joint mechanisms are seen to perform differently for different fiber geometries. This was discussed in Chapter 4 and the differences were attributed to the effects of core diameter and NA. To further complicate the situation the specified parametric tolerances differ for the different fiber geometries. Therefore to estimate the random connection effect it is necessary to consider the following issues:

- joint performance for a specified fiber geometry using the ideal fiber model;
- the impact of interference-fit model to the particular type of joint to be assessed;
- the impact of basic parametric mismatches for the specified fiber geometry.

The following section discusses in some detail the calculation of the various effects for popular fiber geometries and summarizes the state of connection technology with reference to fiber production standards.

The introduction of optical fiber within joint mechanisms

For the purposes of this section only two fiber geometries are considered: 8/125 μm single mode and 50/125 μm multimode. Obviously any number of multimode geometries could be investigated but the greater proportion of multimode fibers installed are either 50/125 μm or 62.5/125 μm. The ideal fiber model joint losses in 62.5 μm, 0.275 NA fiber are approximately the same as those in 50 μm, 0.20 fibers because the impact of the smaller core tends to be balanced by the higher NA.

Random mated insertion loss for 50/125 μm joints

In Chapter 4 it was demonstrated that if the paper specification for 50/125 μm fiber was used then the fiber-related misalignment losses calculated could be:

(a)	core diameter	53–47 μm	1.02 dB
(b)	numerical aperture	0.215–0.185	1.31 dB
(c)	core eccentricity	±2 μm (= 4 μm)	0.46 dB
		Total	2.79 dB

This requires the unlucky installer finding one fiber which features all three parameters at one end of the tolerance spectrum and then jointing it to another which features all three parameters at the opposite end of the tolerance spectrum.

Using a simple statistical approach it is obvious that the probability of this is very small. However, the chance of having some mismatch is quite high. It is useful to have a simple method of assessing the loss on a statistical basis.

Assuming that each parameter varies according to a Gaussian distribution, then the fiber specification can be rewritten in the following form:

Parameter	*Nominal value*	*Standard deviation*
core diameter	50 μm	1 μm
numerical aperture	0.2	0.005
core eccentricity	0 μm	0.7 μm

The combination of all three parameters to create an overall misalignment factor can be treated quite simply by manipulating the standard deviations which represent the probability of achieving a given misalignment. In this way it is found that the insertion loss histogram produced by basic parametric mismatch behaves as if the fiber specifications were as follows:

Parameter	*Nominal value*	*Standard deviation*
core diameter	50 μm	0.2 μm
numerical aperture	0.2	0.0009
core eccentricity	0 μm	0.14 μm

This suggests the following mismatch losses:

(a)	core diameter	50.6–49.4 μm	0.21 dB
(b)	numerical aperture	0.2027–0.1973	0.23 dB
(c)	core eccentricity	±2 μm (= 4 μm)	0.09 dB
		Total	0.53 dB

This calculation shows that a 50/125 μm fiber which fully explores the written specification may, under conditions of total cladding alignment, produce as much as 0.5 dB insertion loss (within three standard deviations). This is quite a high loss despite being considerably better than the ultimate random mated worst case loss of 2.79 dB discussed above. Luckily fiber production methods are such that the full specification limits are not explored and in practice the actual tolerances are perhaps some 40% better than specification. This results in the following 'effective' tolerances and losses:

Parameter	Nominal value	Standard deviation
core diameter	50 μm	0.12 μm
numerical aperture	0.2	0.0005
core eccentricity	0 μm	0.10 μm

This suggests the following mismatch losses:

(a) core diameter	50.4–49.6 μm		0.14 dB
(b) numerical aperture	0.2015–0.1985		0.13 dB
(c) core eccentricity	±2 μm (= 4 μm)		0.04 dB
	Total		0.31 dB

So it would appear that any type of joint mechanism, be it a fusion splice, a mechanical splice or a demountable connector, must have a random mated worst case insertion loss of at least 0.31 dB. This is the base figure to which must be added the other losses to produce the true figure for that joint mechanism.

Manufacturing tolerances for fiber and connectors continue to improve, and most installers would expect to see averages of around 0.2 dB insertion loss across multimode connectors and splices. However, the ISO 11801, EN 50173 and TIA/EIA 568 standards all use the network design values of 0.75 dB per connector and 0.3 dB per splice, regardless of fiber type. Many people see these figures as very conservative, but calculations, and experience, have shown these figures to be realistic worst case figures, and adherence to them in cable network design will avoid 'close-to-the-edge' design problems and ensure operation of all LAN protocols.

Random mated insertion loss for 8/125 μm joints

This is a slightly more complex situation than that seen in the case of multimode fiber. So far in this book single mode fiber has been discussed as having an 8 μm diameter core, whereas measurements of light emitted from a single mode fiber appears to suggest that the light travels along the fiber as if it had a core diameter of 10 μm. This is called the mode field diameter. All the subsequent analyses carried out upon the potential misalignments within single mode joints use the 10 μm value for core diameter.

Because of the step index core structure the NA of single mode fibers is better controlled as is the mode field diameter. The eccentricity of the core within the cladding is limited to 0.5 μm and this factor is responsible for most of the basic parametric mismatch loss. Most manufacturers claim average insertion losses of 0.1 to 0.15 dB for their single mode connectors, but once again note that the recommended design allowance from the standards is 0.75 dB.

Joint mechanisms: relative cladding diameter alignment

Fusion splicing

The use of fusion splices as a jointing technique perhaps is the most efficient method of joining two separate fibers using relative cladding diameter alignment.

A V-groove within a machined metal block is used to align the cladding surfaces of two separate fibers. If the cladding diameters of the two fibers are the same, then the misalignment losses will be parametric in nature. If the cladding diameters are different, then additional core eccentricity misalignment must be considered.

Cladding diameter for most 125 μm fibers can be up to ±3 μm, which means that the maximum additional core–core misalignment is 3 μm. In practice the production of 125 μm fiber results in the distribution shown below:

Cladding diameter (μm)	Population (%)
124–6	70
123.5–6.5	97
123–7	99.5

As a result the use of V-groove alignment may incur an additional core misalignment limited to a further 3 μm (0.34 dB) with 97% of all results better than 0.2 dB.

Multimode fusion splicing

For multimode fibers the jigs used for alignment tend to be fixed in both X- and Y-axes as shown in Figure 5.8. Fixed V-grooves as shown above can lead to perhaps 0.34 dB misalignment loss due to cladding tolerances in addition to the 0.31 dB due to basic parametric mismatch. However, the complex interaction of surface tension effects with the melt of the

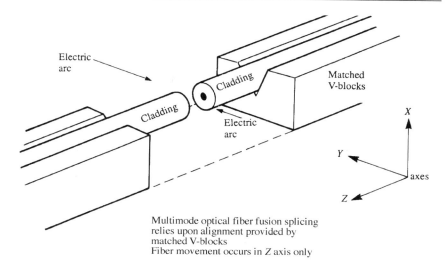

Figure 5.8 *Multimode fusion splicing*

two fusing fibers tends to reduce the true cladding misalignment. This restricts the random mated worst case insertion loss to approximately 0.5 dB.

Single-mode fusion splicing

The equipment used to fusion splice single mode 8/125 μm fibers tends to be larger and more expensive than its multimode counterparts. This is because the alignment technique includes V-grooves which are driven in both the X- and Y-axes. This enables full alignment of the cladding, which overcomes the potential losses resulting from the 3 μm (maximum) misalignment discussed above which would be totally unacceptable for a fiber with a core diameter of 8 μm (or mode field diameter of 10 μm).

Using this equipment the random mated worst case insertion loss is restricted to 0.5 dB, the same as for multimode fibers (50/125 μm and 62.5/125 μm).

A local injection detection system (LIDS) is used to locally inject light into the fiber core in the vicinity of a prepared splice. A detection system situated at the other side of the prepared splice measures the light transmitted prior to fusion. These systems, which inject and detect the light by the application of macrobending, optimise the transmission by manipulating the fibers using the driven V-grooves. In theory this removes the core eccentricity content within the basic parametric mismatch factor. However, in this instance, the surface tension and viscosity effects act upon the entire fiber, which can sometimes negate the advantages of LIDS based equipment.

Mechanical splices

Where the effect of fusion processes tends to reduce the losses due to cladding diameter tolerances the V-groove based mechanical splices must accept any such variation and are therefore inherently more lossy.

Multimode mechanical splices therefore have a random mated worst case insertion loss of the full 0.65 dB (0.31 dB parametric + 0.34 dB cladding) to which must be added any Fresnel loss (where index matching gel is not used) and end-face separation effects. The best multimode mechanical fiber splices, using index matching gel, achieve a final figure of some 0.2 dB average. Mechanical splices are not usually used with single mode fiber, where insertion loss could be up to 1 dB.

Joint mechanisms: absolute cladding diameter alignment

Demountable connectors

In the previous section it was shown that the relative cladding diameter alignment type of joint could create insertion loss figures based upon random mated worst case conditions as detailed below.

	Single mode	multimode
Fusion splice	8/125 0.5 dB	50/125 0.5 dB
Mechanical splice		
with index match	1.0 dB	0.8 dB
with optimization	0.5 dB	0.5 dB

These insertion losses are, in general, higher than would be quoted by the manufacturers of the specific joint components but nevertheless they are the figures that should be used in subsequent network specifications unless certain steps are used to overcome them. Note that ISO, CENELEC and TIA/EIA standards require the use of 0.75 dB insertion loss for connectors in order to give a realistic system average.

These losses represent the worst case figures using the best jointing technology, but the need for network flexibility has led to the widespread

Practical aspects of connection technology 101

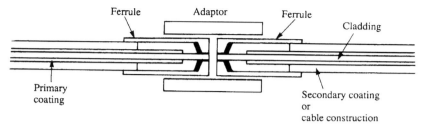

Figure 5.9 *Basic demountable joint*

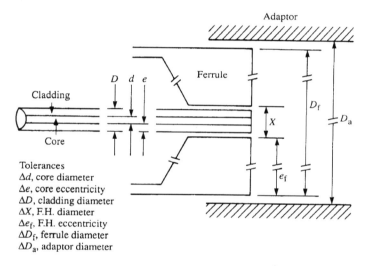

Figure 5.10 *Demountable joint misalignment errors*

use of demountable connectors.

In their basic format demountable connectors are probably the best example of joints using absolute cladding diameter alignment techniques. Figures 5.9 and 5.10 show a typical demountable joint and the misalignment errors that can be built up. Absolute cladding diameter alignment occurs in joints where the fiber is held within a ferrule bore of a given diameter. For example, if the ferrule hole diameter is 128 μm then, unless steps are taken to minimize misalignment, a fiber of diameter 124 μm can easily be misaligned by 2 μm against a fiber of diameter 128 μm. This element of misalignment is additional to the basic parametric factors discussed earlier in this chapter and in some cases can dominate the final insertion loss calculation.

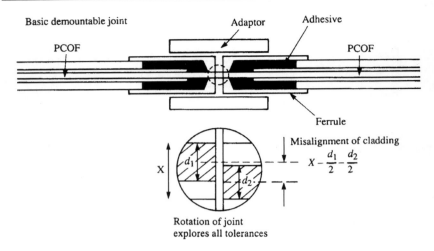

Figure 5.11 Absolute misalignment

Basic connector design

The most basic design of demountable connector comprises a ferrule of hole diameter 128–129 μm into which the fiber is bonded using an appropriate adhesive as in Figure 5.11. In this most fundamental form the ferrules can be rotated through 360° within an alignment tube or adaptor.

This rotation allows any eccentricity to be fully explored as is discussed below.

In addition a gap of perhaps 5–10 μm is incorporated between the ferrule faces. This gap is intended to prevent damage to the fiber ends but has the disadvantage that it incurs both Fresnel loss and loss due to separation effects as discussed in Chapter 4. Also the gap creates reflections, which limits the achievable return loss to approximately 11 dB.

It is worth while to review the basic joint losses observed in practice as opposed to those provided in manufacturers' data, which may have been generated using the interference-fit model or similar distortions. Obviously these losses operate in tandem with losses caused by basic parametric mismatches.

The ferrule hole diameter (FHD) is of considerable importance since the cladding misalignment achieved when jointing a fiber of cladding diameter d_1, to another with cladding diameter d_2 in an FHD of X will be seen from equation (5.2), in the worst case:

$$\text{cladding misalignment} = X - (d_1 + d_2)/2 \tag{5.2}$$

For a fixed FHD the impact of using fibers with diameters at the lower end of the specified tolerance can be quite severe. For instance, for FHD of 129 µm the misalignment for two 122 µm fibers can be as much as 7 µm, which could result in a power loss of 0.85 dB for 50 µm core fibers and total loss for single mode fibers.

Two factors work against such losses being encountered. First the specification for cladding diameter is rarely fully explored, with 97% of all fiber lying in the range 123.5–126.5 µm. Taking due account of these factors limits the misalignment loss to approximately 3 µm, corresponding to a loss of some 0.30 dB for a 50 µm core fiber. Although this figure is considerably lower than the 1.85 dB shown above it is nevertheless much greater than would be predicted using the interference-fit model (0 dB).

Second, the professional termination of the connectors involves filling the ferrule with an adhesive, normally some type of epoxy resin. The fiber is guided through the adhesive and passes through the hole in the ferrule end face. For smaller fibers the adhesive serves to assist in the centralization of the fiber in the hole, thereby reducing the eccentricity.

One of the first optical connectors to be used in data communications was the SMA. This basic demountable connector exhibits random mated worst case insertion losses between 1.3 dB and 2.0 dB dependent upon the quality of the connector components themselves.

Keyed connectors

The fact that the ferrules within basic demountable connectors can be rotated through a full 360° implies that the impact of core eccentricity (due to fiber manufacturing tolerance) and cladding-based misalignment (due to fixed FHD) will inevitably surface in the form of rotational variations in insertion loss. This results in poor repeatability.

This undesirable rotational degree of freedom has since led to the introduction of keyed connectors, e.g. keyed or bayonet-mount connectors such as the ST and FC, which define the orientation of the ferrule face against another. This prevents rotation and ensures good repeatability. However, this repeatability is only guaranteed in a given joint since the inherent cladding-based misalignments are still present.

An example would be that ferrule A against ferrule B may achieve 0.5 dB insertion loss in a highly repeatable and stable fashion. Also ferrule C against ferrule D may achieve the same performance; however, the difficulty arises when ferrule A is measured against ferrule C or B against D. In these circumstances the insertion loss is unpredictable (but repeatable) although it will lie with the bounds of the limits of the random mated worst case calculation for the joint design.

As a result keyed connectors do not necessarily give better results for insertion loss but are perceived to perform in a more repeatable way; however, this is only true for a given joint.

6 Connectors and joints, alternatives and applications

Introduction

Chapters 4 and 5 have concentrated upon the losses generated within joints due to the fibers themselves, the type of alignment and also the quality of the components used.

The conclusions so far are that fusion splice techniques offer the best opportunity for low-loss connection whereas demountable connectors, while offering major advantages in terms of flexibility, cannot achieve these low levels of loss. In addition it has been repeatedly stated that joint performance must take account of the fiber, the alignment technique and the mechanics of the joint not as three separate unrelated issues but rather as three component parts of an integrated structure. Finally it is believed that the best joint mechanisms, whether fusion splice or demountable connectors, are now reaching their ultimate performance and advances in fiber tolerancing will be required before any further improvements can be made.

The reason that the 'joint' has been the focus of so much attention is that it frequently is the foundation of faulty design, poor installation and, eventually, network failure.

Inaccurate assessment of potential joint losses at the design stage can radically affect the flexibility of a network once installed. In the most dire situations communication may be impossible. Poor installation of joints (whether fusion spliced, mechanical spliced or demountably connected) can be a source of network problems characterized by variable link performance and seemingly random failures. Mishandling of joints may contribute to network failure – it is a fact that 98% of all fiber-related failures occur within 3 metres of the transmission equipment (e.g. demountable connectors at the equipment or within patching facilities).

For all these reasons the joint mechanisms used should be the focus of critical scrutiny. They are certainly one of the main inspection issues during cabling installation. This chapter reviews the joint market and examines the suitability of various joints for given environments and applications. The processing of the joints is discussed and the alternatives are assessed technically and financially.

Splice joints

The competition between fusion and mechanical splicing has always been fierce and will continue to be so. It is based upon personal views on issues such as experience, perceptions of cost and, rather obviously, performance.

Fusion splice joints are relatively easy to produce requiring few special hand tools but unfortunately necessitating the use of comparatively expensive optical fiber fusion splicers (OFFS) which can represent a significant capital outlay. That being said the consumables used during splicing are of negligible cost.

On the other hand mechanical splice techniques, whilst again requiring few hand tools, do not require such a large initial expense. The primary disadvantage is that mechanical splices are precision mechanical components and as a result are not cheap (the joints in some cases costing more than the equivalent demountable connector). Also the amount of test or optimization equipment needed must be considered and its cost amortized in some sensible fashion.

Cost analysis of fusion splice jointing

Note: The following examples of costs are given in British pounds. A rough equivalent conversion rate is $1.5 US to the pound and 1.6 Euros to the pound.

The cost of producing a single fusion splice joint is a function of the capital outlay on the equipment required together with the cost of the labour involved in the completion of the joint to the desired specification.

The tooling necessary to undertake fusion splicing is shown in Table 6.1. It will be immediately noticed that there are few specialist fiber optic tools within this list. Indeed the majority of hand tools required are standard copper cabling devices such as cable strippers. As a result the true investment in optical fiber fusion splicing is limited to that shown in Column 3 of Table 6.1.

There is a large difference between the investment required to fusion splice single mode and multimode fibers. This is due to the need to procure higher-quality cleaving tools and more complex fusion splicing equipment for the small core fiber geometry. For the purposes of this analysis it will be assumed that multimode technology is to be adopted and therefore the total cost of the specialist tools and splicer will be approximately £4800.

The method of amortizing this is open to question; however, as installation costs are normally taken on a per-day basis it is sensible to assess the cost of ownership of this equipment as follows.

Table 6.1 *Optical fiber fusion splicing tool kit*

Description	Cost (£) Multimode	Single mode
Fusion splicer	4000	9000
Specialist cable preparation tools	500	500
Fiber cleaving tool	300	900
Total	4800	10 400
Initial cost of equipment (multimode)	= £4800	
Interest on purchase price	= £480 (per annum)	
Depreciation	= £1600 (per annum)	
Repair/maintenance	= £400	
Annualized cost of ownership	£2480 (per annum)	
Operating analysis		
Total no. of working days	= 228	
Maximum no. of on-site days	= 200	
Reductions for calibration/maintenance	= 5 days	

This operating analysis suggests that the maximum number of days that the equipment will experience field use will be 195. Based upon this analysis the cost of ownership is approximately £13 per day.

Obviously for a less than fully occupied team of installers the cost of ownership will be correspondingly greater as is shown below.

Projected on-site days	Cost per day
195	£13
150	£17
100	£25
50	£50

The above analysis certainly suggests that to fusion splice as a standard jointing mechanism can be financially justified only if the installer intends to fully utilize the equipment. Essentially the calculation shows that the cost of ownership increases dramatically as the overall usage drops. To the cost of ownership must be added the unit cost of labour involved in the completion of a successful joint.

In an external installation environment with telecommunications grade cables the jointing process is accompanied by a large amount of preparation and testing (see later in the book). As a result the actual number of fusion splices undertaken by an installation team is unlikely to rise above 16 per day. Fusion splicing factory-made tails to indoor cable is much simpler and quicker and 40 per day are realizable.

The costs of providing a person to site for the purposes of installation are very much dependent upon the environment in which they are expected to operate; however, it is unlikely to be less than £250 per day.

The cost of producing a splice is therefore:
Material cost (protection sleeve) = £0.90
Unit labour cost = £6.25–16.00
Unit cost of ownership = £0.33–0.80

The basic cost of a multimode splice, installed by a well-trained operator working 80–90% of maximum annual capacity is therefore between £7.50 and £17.70. If we insert the figures for single mode equipment, and presume that most of the use would be in external plant, then the cost per splice comes to around £19–20 each.

Cost analysis of mechanical splice jointing

In this section it is assumed that the mechanical joint is a basic device which can be optimized to achieve performance in many cases comparable with a fusion splice joint (assuming that index matching materials are used). Table 6.2 shows the necessary tooling. The mechanism for optimization may be via an optical time domain reflectometer (see Chapter 12) which would be required for final system characterization of any type of

Table 6.2 *Mechanical splicing tool kit*

Description	Cost (£) Multimode	Single mode
Alignment jig	1500	2500
Specialist cable preparation tools	500	500
Fiber cleaving tool	300	900
Cleaving tool	300	900
Microscope	300	300
Total	2900	5200

joint and cannot be solely allocated to the tooling for mechanical splice joints. It is therefore not considered as part of the tooling list.

For the most basic mechanical splice joint the specialist tool kit is estimated to cost approximately £2900. The analysis undertaken in the previous section can be repeated as follows:

Initial cost of equipment	= £2900
Interest on purchase price	= £290 (per annum)
Depreciation	= £960 (per annum)
Repair/maintenance	= £200
Annualized cost of ownership	£1450 (per annum)
Operating analysis	
Total no. of working days	= 228
Maximum no. of on-site days	= 200
Reductions for calibration	= nil

This operating analysis suggests that the maximum number of days that the equipment will experience field use will be 200. Based upon this analysis the cost of ownership cannot be less than approximately £7.25 per day.

Obviously for a less than fully occupied team of installers the cost of ownership will be correspondingly greater, as is shown below.

Projected on-site days	Cost per day (£)
200	7.25
150	9.65
100	14.50
50	29.00

As in the previous section the jointing process is accompanied by a large amount of preparation and testing (see later in the book). Although the mechanical splice joints are frequently advertised as having timing advantages over their fusion splice counterparts, it is felt that the benefit, as will be experienced in the field, will be relatively small, and the number of joints completed to a defined specification is unlikely to rise above 20 to 50 per day, depending upon the environment already described.

The cost of producing a splice is therefore:

Material cost (mechanical joint)	= £8.00
Unit labour cost	= £5.0–12.50
Unit cost of ownership	= £0.15–0.36

The basic cost of a multimode mechanical splice, installed by a well-trained operator working 80–90% of maximum annual capacity is therefore between £13.50 and £21.00. Single mode prices would not be much different.

The two sets of costs are shown below.

Joint type	Internal splicing	External cable plant
Fusion	£7.50	£17.70
Mechanical	£13.50	£21.00

These figures have been generated for multimode fiber jointing. The figures for single mode technology will obviously be greater; however, the trend will be the same. There are two important conclusions to be drawn from the analyses above, as follows:

(1) An organization intending to undertake irregular jointing for purposes of either installation or repair of fiber optic cables should seriously consider the mechanical splice option provided that the training given enables joints to be made within specification.
(2) The cost of ownership of the equipment necessary for either fusion or mechanical splicing is a small proportion of the labour cost involved in producing a joint to an agreed specification. The labour cost is governed by factors completely outside the realm of fiber optics such as salaries, travel and accommodation expenses and the rate at which the tasks can be undertaken. The latter is governed by the environment in which the work is to be carried out, the cable designs and the position and type of joint enclosures rather than the particular jointing technique or testing requirements.

Recommendations on jointing mechanisms

The preceding section suggested that fusion splicing offered cost advantages where usage was forecast to be regular. Mechanical splicing, however, showed some advantage if the jointing tasks were likely to be intermittent.

Before any final decision is made, the skill factors must also be taken into account. In both cases the most skilled task is the cleaving of the fiber – that is the achievement of a square, defect-free fiber end prior to further processing. The process of cleaving a fiber is relatively straightforward with practice and to assist in increasing the 'first-time success' rate various specialist tools have been developed.

Nevertheless, acceptable fusion or mechanical splices depend upon the quality of the cleave, and much time can be wasted by attempting to produce splices from fibers with inadequate cleaved ends.

For this reason a method which incorporates inspection of the cleaved ends as part of the jointing process has marked advantages for the beginner or intermittent user. Optical fiber fusion splicers tend to feature integral microscopes which are used both for monitoring fiber alignment and to inspect the fiber ends. As a result the throughput of good joints can be higher than that for mechanical splices, where lack of confidence can create low cleave yields which in turn result in poor joints which have to be remade. This repeated use of the mechanical splice component can result in damage which increases the material cost of the joint in addition to the labour cost impact.

To summarize, the lower capital cost of a mechanical splice option can disguise a high unit production cost — this can be further affected by poor yields at the unit material cost level if cleave yields are poor (due perhaps to infrequent operation). This must be balanced by the higher capital cost of fusion splicing equipment which must be regularly used to justify this outlay. Perhaps the obvious solution is to lease, hire or rent the capital equipment needed for fusion splicing only when it is needed (provided that a familiarization period is included prior to formal use).

As a general rule of thumb, any installer doing more than about 200 splices per year should seriously consider purchase or lease of a fusion splicer.

Demountable connectors

Since the earliest days of fiber optics the demountable connector has been somewhat of a poor relation to the high-technology components such as the optical fiber itself, the LED and laser devices and their respective detectors. The need for demountable connectors was clear in terms of flexibility, repairability and the more obvious need to launch and receive light into and from the fiber. However, the mechanical properties of the demountable connectors such as accurate hole dimensions, tight tolerances on hole concentricity and ferrule circularity were not easily achieved and as a result the developments have been slow and, until recently, standards have been difficult to set.

As the early application of optical fiber was for telecommunications, the standardization reflected the national preferences of the PTT organizations. For instance, the UK telecommunications groups defined approved connector types for both multimode and single mode applications. Unfortunately the rapid move away from multimode transmission within telecommunications led to little desire for technical improvements for multimode connector styles and as a result the general level of standardization is correspondingly lower than for single mode variants.

The single mode demountable connectors such as the NTT designs (FC and FC/PC) are seen in virtually all world markets and are supplied by indigenous manufacturers to a tightly controlled dimensional format giving high levels of intermateability and interoperability.

The multimode market, left behind by the telecommunications applications, was forced to develop new designs and improved performance

Table 6.3 *Common demountable connector styles*

Usual application	Common designation
Multimode applications	ST
	SC
	MT-RJ
	SG
	LC
Single mode applications	SC
	FC-PC
	LC

levels without national approvals (and the large usage that implied). As a result the range of designs available increased to an unacceptable degree and the multimode connector was repeatedly criticized for a lack of standards. During the 1990s the data communications market has matured very rapidly and rationalization of designs and manufacturers has occurred.

The SC and ST connector styles have accounted for some 98% of all multimode connector usage within Europe and the USA. Until the late 1990s the ST and SC retained their dominance in the structured cabling, multimode environment, and the latest standards still give preference to the SC. Many new styles have now been introduced, mostly under the generic heading of SFF, or small form factor, which basically means a multi-fiber optical connector of a similar cross-sectional area to a copper-cable RJ-45 connector. The large and very diverse user base for multi-mode connectors has made the market quite conservative and slow to adopt new designs except where clear advantages can be seen.

Table 6.3 details the currently available styles for both multimode and single mode fiber geometries and these are discussed below.

Basic ferrule designs

In the earliest days of optical fiber jointing the sophisticated manufacturing techniques now associated with demountable connectors were not available. Instead existing components or technology had to be used or modified to provide an acceptable level of insertion loss.

The most obvious method was to use a machined V-groove as an alignment tool and to secure the two fiber ends within metal tubes or ferrules in such a way as to ensure acceptable performance. It is relatively easy to produce V-grooves and tubes to the required tolerances: the difficult task was to align the fibers within the tubes themselves. These basic ferrule designs resorted to watch jewels made from synthetic ruby or sapphire

112 Fiber Optic Cabling

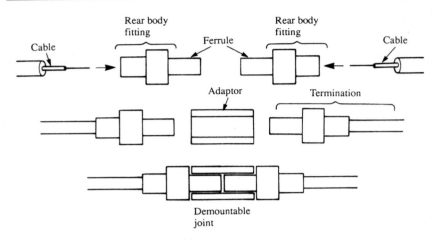

Figure 6.1 Basic demountable joint

which were inserted into the tubes (see Figure 6.1). These jewelled ferrules are still used as methods of connection to test equipment as they are simple to terminate and their performance is limited by the hole diameter, concentricity and fitting tolerances of the jewel within the tube.

The first true demountable connector was the SMA (subminiature 'A') which was loosely based upon the design of the electrical connector of the same name.

It featured an all-metal ferrule together with a rear body which allowed stable connection to the incoming cable. The rear body provided connection to the alignment tool (a threaded tube) by means of a captive nut.

At this time the following basic connection terminology was defined as shown in Figure 6.1.

- *Connector.* The complete assembly attached to the fiber optic cable. In general these are of male configuration requiring a female component to allow the jointing of two connectors. In technical literature the connector may be termed a plug.
- *Adaptor.* The female component used to join two connectors. The adaptor is responsible for providing the alignment of the connectors and the fibers within them. Other terms frequently used are: uniters, couplers, sockets, receptacles.
- *Termination.* The process of attaching a connector to a fiber optic cable element (and also the name given to the completed assembly).

Most fiber optic connectors comprise a ferrule (or in the case of multi-element devices, a number of ferrules) which is responsible for the control of fiber alignment and a rear body which is responsible for the attachment of the connector to the cable. The rear body can be complex in construction and has additional responsibilities for the connection of the connector to the adaptor. The mechanics of the ferrules and rear bodies vary from connector to connector and will be discussed for each connector type. The ferrule is usually made of a ceramic material, because of its hardness and thermal stability, although polymer ferrules are now accepted as being nearly as good. The particular ceramic is known as zirconia (zirconium dioxide ZrO_2) chosen because its hardness and thermal coefficient of expansion is very similar to glass. It also polishes well. To be more specific the material is usually tetragonal zirconia polycrystal, which is zirconia and yttria (yttrium oxide Y_2O_3).

ST fiber optic connectors

Although the biconic connector was approved by the standards bodies, the early 1990s market soon became dominated by the ST, or ST II connector as it is often called, or even sometimes the BFOC/2.5. The ST connector is a true fiber optic connector designed to meet the requirements of low insertion loss with repeatable characteristics in a package which is easy to terminate and test.

The ST is a keyed connector with a bayonet fitting. The ferrule is normally ceramic based (although polymer ferrules are often promoted as having equivalent performance at a lower price) as are virtually all the modern high-performance connectors (multimode and single mode). It is also a spring-loaded connector which means that the ferrule end-face separation is not governed by the adaptor but rather the separation tends to zero due to the spring action within the mated ferrules.

As with all technical advances there are some disadvantages, and the spring-loaded connector is no exception:

Springing methods
There are two methods which can be adopted to produce spring action within the ferrule of a fiber optic connector. The first is sprung body, the second sprung ferrule.
- *Sprung body designs* (see Figure 6.2). When connected into the appropriate adaptor and the cable pulled a spring loaded connector can respond in one of two ways: if the ferrule springing tends to be released (i.e. the ferrule ends separate) then the design features a sprung body. This simply means that the cable is directly linked to the ferrule and the rear body of the connector is spring loaded against the ferrule. This type of springing prevents any stress within the connector but certain users are concerned about interruption of traffic during the application of pull to the terminated cable and prefer to use sprung ferrule designs.

114 *Fiber Optic Cabling*

- *Sprung ferrule designs* (see Figure 6.3). The alternative to a sprung body is a sprung ferrule. In this case there is no movement when the cable is subjected to a pull because the rear body, attached to the adaptor, is directly linked to the cable. The ferrule is therefore spring loaded against the body.

The disadvantage to this approach is that the terminated fiber within the ferrule comes under significant compressive force as the connector is tightened or clipped into place within the adaptor. This compressive force can, under certain conditions, result in large amounts of optical attenuation due to microbending. It is therefore necessary to ensure that the cabling components are relatively free to move and that there are no

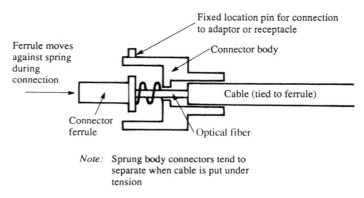

Figure 6.2 *Sprung body connector design*

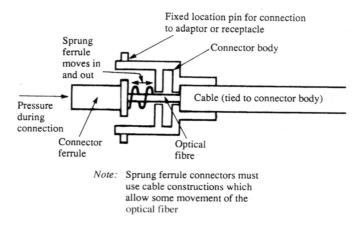

Figure 6.3 *Sprung ferrule connector design*

constrictions (due to tight fitting of cable to connector, for instance). This puts increased responsibility upon the installer in a technical area well beyond the expectations of the average contractor.

The ST connector is also capable of high performance on single mode fiber geometries. The design and manufacturing quality of the adaptors is such that by the use of tight tolerance ferrules (sized ferrule components) a single mode performance equivalent to the best single mode connectors can be achieved.

Other connector designs

The SC and ST dominate the world markets; however, there are a number of other multimode connectors worthy of mention due either to their historic importance or growing influence.

- *SMA*. The first mass produced multimode optical connector for the data communications industry; now obsolete.
- *Biconic*. A sprung connector that was favoured by IBM in the early 1990s, but offered little in the way of price or performance benefits. Now obsolete.
- *FDDI or MIC connector*. A duplex connector in a large body shell designed specifically for the FDDI LAN interface. Now obsolete.
- *IBM ESCON®*. A duplex connector very similar in appearance to the FDDI MIC connector. Only found on particular IBM mainframes, ESCON Directors and look-alikes from other manufacturers. Only applicable now to IBM mainframe installations.
- *MT*. A ribbon or 'mass termination' ferrule developed by NTT specifically for optical ribbon cables. It featured one small rectangular ferrule with 4, 8, 12 or even 16 holes drilled in it. Two alignment pins were responsible for the principal alignment and a spring clip held the two identical halves together. An index matching gel was usually used between the end-faces to improve performance. The MT became popular with PTTs who used high fiber-count ribbon cables.
- *MPO*. The MPO uses the MT ferrule inside a larger, spring-loaded plastic housing. It makes the MPO an easy-to-handle connector for regular patching and cross-connect activity. The MT on its own is seen as a 'demountable splice' and too small for patching duties outside of a patch panel or enclosure.
- *MU*. The MU was introduced by NTT in 1995 as a high density multiway connector (2, 8 and 16) for transmission systems, optical switching and optical backplanes. It is selected for the IEEE P1355 FutureBus Architecture.
- *ST*. The most commonly used connector for data communications throughout the 1990s. Recognized as an alternative to the SC

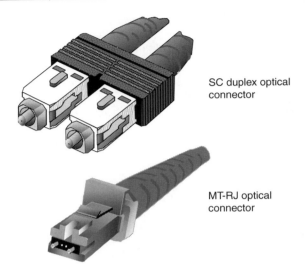

Figure 6.4 *SC duplex and MT-RJ optical connectors*

connector in structured cabling standards ISO 11801 1st edition, EN 50173 1st edition and TIA/EIA 568A. The ST is no longer mentioned in ISO 11801 2nd edition.

- *SC.* (Figure 6.4). Originally the 'Subscriber Connector' from Japan. The SC is a push-fit connector in both multimode and single mode versions. It is the first connector of choice in both editions of ISO 11801 and EN 50173. It can be supplied in simplex (i.e. one connector) or duplex (i.e. two connectors next to each other) format. The international standards refer mainly to the duplex version.
- *FC.* The FC, also from NTT in Japan, has become the industry leading single mode connector. It is more commonly known, however, as the FC-PC, where PC stands for physical contact.
- *Optijack.* The first of the new generation SFF or small form factor connectors. The Fiberjack was introduced by Panduit in 1996. The Fiberjack uses two 2.5 mm ferrules in one single RJ-45 type housing.
- *LC.* (Figure 6.5). A duplex SFF connector from Lucent which used two 1.25 mm ferrules. The LC can be split into two simplex connectors.
- *VF-45.* A ferrule-less duplex connector introduced by 3M, also known generically as the SG or as Volition, when part of 3M's optical cabling system, aimed at lowering cost in the fiber-to-the-desk market.
- *MT-RJ.* (Figure 6.4). Probably the leading contender in the SFF connector market and supported by many manufacturers such as AMP (Tyco), Siecor (Corning) and Fujikura. It is based on the original MT ferrule design.

Figure 6.5 *LC optical connector*

Apart from specialist optical connectors for the military and industrial markets there are still more optical connectors around for the data and telecommunications market. There are the D3 and D4 connectors, the E2000, the LX5, the Mini-Mac from Berg handling up to 18 fibers, and the Mini-MT and SCDC/SCQC both from Siecor (Corning), where DC means duplex and QC means quadruplex, or four, fiber terminations.

Single mode connectors

Single mode connectors have to be made to a higher manufacturing tolerance than multimode connectors, and the use of high-powered communications lasers makes return loss more of an issue within the laser-based telecoms market compared to data communications.

The connector designs that have been more widely adopted than any other are the ceramic-based styles originating from the Japanese, normally under licence from Nippon Telephone and Telegraph (NTT). The Japanese dominate the high-quality ceramic industry so it is no surprise that the connectors using the material as an alignment technique also came from that nation.

There are a number of early Japanese multimode and single mode connectors in use such as the D3 and D4 series; however, the design that has become dominant in the world market is the FC (or FC/PC) and the single mode SC.

The FC connector generally features a ceramic ferrule and is keyed and can be optimized against a master cord. The connector is made by a large number of companies and is available in either sprung ferrule or sprung body formats. The basic FC design was intended to be terminated with a flat fiber/ferrule end-face; however, its use on single mode fiber developed the concept of physical contact and angle polishing.

As has already been mentioned the return loss, the power reflected back into the launch fiber, is related to the fiber end-face separation in the mated joint. In order to minimize this gap the connector is sprung. The spring acts to put the ferrules under compression and does indeed improve the possible performance. Unfortunately termination techniques cannot guarantee face-to-face mating over the entire core area and the PC surface finish to the termination was introduced.

PC is an abbreviation for physical contact and is a successful attempt to profile the fiber end in order to provide deformation of the optical core areas within a joint to the point where no end-face separation exists. In the case of the FC/PC this is achieved by creating a convex surface at the ferrule end (with a radius of approximately 60 mm). Over the 8 μm optical core diameter (or 10 μm mode field diameter) the two mated core areas are therefore compressed (due to the spring action of the connectors) with a resultant improvement in return loss. The FC termination would be expected to achieve the −11 dB calculated in Chapter 4 whereas a PC version would normally achieve −30 dB. This level of improvement is of value to high-speed laser systems where reflections must be kept to a minimum. Flat polished ferrules, with an inevitable air gap, versus spherical polished, physical contact ferrules are shown in Figure 6.6.

There are two methods of achieving a PC termination:

- The ferrule starts out in a flat face format, the fiber is terminated and then the terminated connector is profiled. This is normally undertaken using a purpose-built polishing machine fitted with concave polishing pads. Diamond polishing pastes are used to create the profile in the ceramic ferrule.
- The ferrule is pre-profiled and therefore is always a PC ferrule. If polishing takes place on a hard surface (as is normal for flat faced terminations) then the final finish tends to be FC. Alternatively polishing on a soft material allows the fiber to take on the form of the ferrule's surface profile, hence becoming a PC termination.

The FC/PC connector has become a world standard demountable joint for single mode applications since it offers the telecommunication industry the benefits of low insertion loss, good return loss, repeatable performance (due to keying) and optimization. As the specification of optical fibers improves some of these features may become less necessary (indeed optimization is not always required to meet telecommunications

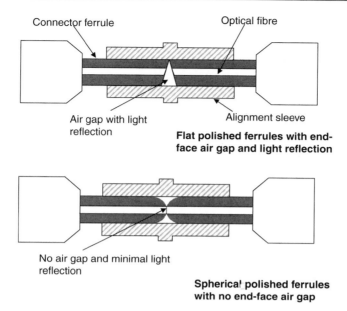

Figure 6.6 *Flat polished and spherical polished ferrules*

requirements) and as a result other connectors have been accepted as being suitable for single mode applications. Examples of these are the ST, SC, LC and MT.

The aspect of physical contact itself has developed into four categories, shown in Table 6.4.

End-face geometry is defined by the following parameters:[1]

- *Fiber height.* The fiber undercut or protrusion with respect to the ferrule end-face.
- *Apex offset.* The distance between the peak of the sphere that best fits the ferrule end-face and the axis of the fiber.

Table 6.4 *Physical contact categories*

Category of PC	Symbol	Multimode Max return loss dB	Single mode Max return loss dB
Physical contact	PC	20	30
Super physical contact	SPC	N/A	40
Ultra physical contact	UPC	N/A	55
Angled physical contact	APC	N/A	60

120 Fiber Optic Cabling

Table 6.5 *End-face geometry parameters*

Parameter	Symbol	Value
Apex offset	a	<50 µm
Radius of curvature	R	7–25 mm
Ferrule undercut	x	125 nm when R is between 7 and 10 mm
		50–125 nm when R is between 10 and 25 mm
Ferrule protrusion	y	<50 nm

- Radius of curvature of the ferrule end-face.
- Contact force.
- Material used in the ferrule, fiber and epoxy.

Table 6.5 and Figure 6.7 demonstrate the meaning and typical values of these geometries, and are based on Telcordia[2] (formerly Bellcore) GR-326-CORE issue 3.

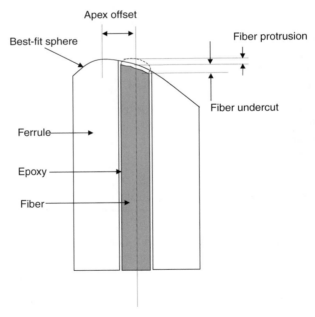

Figure 6.7 *End-face geometry parameters*

Standards and optical connectors

The subject of optical connectors started the 1990s with no standards basis or coverage at all, but starts the new millennium with a host of standards covering every aspect of mechanical and optical specifications. The standards, however, continue to evolve and mature, as does the rest of the subject.

Optical connectors are mostly specified by two bodies, the TIA in America (Telecommunications Industry Association, www.TIAonline.org) and the IEC (International Electrotechnical Commission, www.IEC.ch).

American standards

The TIA writes optical test procedures, the FOTP series (*ANSI/EIA/TIA-455-A-1991 Standard Test procedures for fiber optic fibers, cables and transducers, sensors, connecting and terminating devices, and other fiber optic components*). Note that in the USA standards are usually published under the auspices of ANSI, or American National Standards Institute, whereas the detailed work, in this field, is done by the TIA and EIA, Electrical Industries Alliance.

The TIA also have FOCIS, Fiber Optic Connector Intermateability Standard, published as TIA/EIA-604. This standard defines the mechanical parameters of different optical connectors, adaptors and receptacles. The published list of FOCIS connectors (February 2001) is:

TIA/EIA 604-1	April 1996	Biconic connector
TIA/EIA 604-2	November 1997	ST
TIA/EIA 604-3	August 1997	SC
TIA/EIA 604-4	August 1997	FC
TIA/EIA 604-5	November 1999	MPO
TIA/EIA 604-6	March 1999	Fiber Jack
TIA/EIA 604-7	January 1999	SG (VF-45)
TIA/EIA 604-10	October 1999	LC
TIA/EIA 604-12	September 2000	MT-RJ

The missing numbers, e.g. FOCIS 11, are for connector proposals which are still in discussion, in draft, or in ballot. The proposal for FOCIS 11 was the Siecor SCDC/SCQC connector.

The connector and test specifications are then called up in structured cabling standards. For example, TIA/EIA-568-B3, Commercial Building Telecommunications Cabling Standard part 3, Optical Fiber Cabling Components, states:

> Various connector designs may be used provided that the connector design satisfies the performance requirements specified

within annex A *(of TIA/EIA 568-B-3, which in turn calls up numerous FOTP test standards from ANSI/EIA/TIA-455)*. These connector designs shall meet the requirements of the corresponding TIA FOCIS document. The duplex SC... referred to as the 568SC...*(later referenced as ANSI/TIA/EIA-604-3, 3P-0-2-1-1-0 for single mode, 3P-0-2-1-4-0 for multimode and 3A-2-1-0 for adaptors)* is referenced for illustrative purposes in this Standard.

(Author's comments in italics)

In other words, any connector is allowed in the American TIA/EIA 568B standard as long as it is on the FOCIS TIA/EIA 604 list and meets all the optical and mechanical specifications of the referenced FOTP tests.

International standards

The IEC has many standards relevant to optical connectors, produced by the subcommittee known as IEC SC86B.

IEC 61300 covers standard environmental and mechanical test methods
IEC 61754 and 60874 provide mechanical and dimensional information
IEC 61753 provides tests and measurements with pass/fail criteria
IEC 62005 addresses aspects of reliability

Some particular standards of greatest relevance are:

IEC 60874-7	1993	FC
IEC 60874-10	1992	BFOC/2.5 (ST)
IEC 60874-10-1	1997	BFOC/2.5 multimode
IEC 60874-10-2	1997	BFOC/2.5 single mode
IEC 60874-10-3	1997	BFOC/2.5 single mode and multimode
IEC 60874-14	1993	SC
IEC 60874-14-1	1997	SC/PC multimode
IEC 60874-14-2	1997	SC/PC tuned single mode
IEC 60874-14-3	1997	SC single mode simplex adaptor
IEC 60874-14-4	1997	SC multimode simplex adaptor
IEC 60874-14-5	1997	SC/PC untuned single mode
IEC 60874-14-6	1997	SC APC 9° untuned single mode
IEC 60874-14-7	1997	SC APC 9° tuned single mode
IEC 60874-14-9	1999	SC APC 8° tuned single mode
IEC 60874-14-10	1999	SC APC 8° untuned single mode
IEC 60874-16	1994	MT
IEC 60874-19	1995	SC Duplex
IEC 60874-19-1	1999	SC-PC Duplex multimode patch
IEC 60874-19-2	1999	SC Duplex single mode adaptor
IEC 60874-19-3	1999	SC Duplex multimode adaptor

IEC 61754-2	1996	BFOC/2.5 (ST)
IEC 61754-4	2000	SC
IEC 61754-5	1996	MT
IEC 61754-6	1997	MU
IEC 61754-7	2000	MPO
IEC 61754-13	1999	FC-PC
IEC 61754-18	draft	MT-RJ
IEC 61754-19	draft	SG
IEC 61754-20	draft	LC

ISO 11801 2nd edition 2002, *Interconnection of Information Technology Equipment*, calls for optical cables in the work area to be terminated with a duplex SC connector (SC-D) that complies with IEC 60874-19-1. The standard is a little vague about what connector is allowed in other positions away from the work area and to be safe, details a mechanical and optical performance range in its Table 69, which invokes mechanical and optical specifications derived from IEC 60793-2, IEC 60794-2, IEC 60874-1, IEC 61300-3, IEC 61073-1 and IEC 60874-1.

The figures can be summarized as:

- mechanical endurance   > 500 cycles
- insertion loss, as a connector   0.75 dB
- insertion loss, as a splice   0.3 dB
- return loss, multimode   20 dB
- return loss, single mode   35 dB

Insertion loss should be measured under overfilled launch conditions.

ISO 11801 remarks that when density of connections is important, e.g. at campus, building or floor distributors (i.e. patch panels), then SFF connectors that meet an approved IEC standard can be used.

The SC duplex is shown as a crossover connection, i.e. A to B and B to A, see Figure 6.8.

Colour coding is also mentioned in the standard for SC duplex optical connectors:

- multimode 50/125 and 62.5/125   beige
- single mode PC   blue
- single mode APC   green

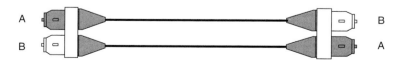

Figure 6.8 *SC duplex patchcord*

Other standards

In America there are the Telcordia standards, formerly known as Bellcore, which also offer component specifications. Different LAN standards also often state which connectors are expected to form the physical interface on that LAN. Gigabit Ethernet, IEEE 802.3z specifies the SC duplex, whereas Fiber Channel and the ATM Forum recognize most of the SFF connectors.

Termination: the attachment of a fiber optic connector to a cable

The process of attaching a fiber optic connector to a cable is frequently underestimated. Perhaps because the action of terminating the most basic copper connectors is so simple (achieving a metal-to-metal contact) it is common, even within the fiber optics industry, for the termination process to be overlooked and regarded as an unimportant issue. As a result a great deal of damage has been done both to installed systems and to the reputation of the technology.

A fiber optic connector is a collection of finely toleranced mechanical components. A fiber optic cable is, as will be seen in Chapter 7, a combination of optical fiber, strength members and plastic or metal sheath materials.

The basic performance of the completed joint is dependent upon the optical characteristics and tolerances of the fiber together with the mechanical tolerances of the connectors and adaptors. However, the termination process can drastically alter the finished performance of the joint. The process affects the outcome at two levels. First the opto-mechanical stresses applied to the fiber within or around the connector can increase the insertion loss of the joint by a factor of 10 000 (40 dB). Second, the physical defects in the termination can, if undetected at the inspection stage, damage other mated components and, in the most extreme cases, destroy remotely connected devices (lasers in the case of high levels of reflected signal).

The act of terminating a fiber optic cable with a connector therefore carries with it a significant responsibility and this section reviews the options open to the installer wishing to produce a network which includes demountable connectors.

The role of a termination

A fiber optic cable can take many forms, from a single element secondary coated fiber without any additional tensile strength to a multi-element structure containing fibers surrounded by strength members. Both can be terminated in the correct style of fiber connector. Regardless of the type of components used the role of a termination is to:

- provide a mechanically stable fiber/ferrule structure which, once completed, shall not undergo failure, thereby damaging other components;
- provide a mechanically stable cable–connector structure which serves to achieve the desired tensile strength for the components chosen;
- induce no additional stresses within the fiber structure, thereby enabling the connector to perform as it was designed to do.

These three prerequisites of a correctly processed termination are valid for all terminations other than those used for temporary purposes such as testing unterminated cables. All three are equally important and their importance is discussed below.

Mechanical stability

The demountable connector may be subject, by its very nature, to frequent handling by both skilled and inexperienced personnel and as a result it is also the component most likely to fail within the fiber optic network.

It is necessary therefore to provide the termination with sufficient tensile, and torsional (or twisting) strength to ensure a reasonable life expectancy. However, the strength of the termination is not simply the figure quoted by the connector manufacturer. That figure will have been measured on a specific cable using a specific type of strength member (where relevant). A different cable construction may not achieve the quoted tensile strength. Therefore a termination cannot be seen as the sum of two specifications: in many cases the tensile strength achieved may be significantly below those quoted by the connector and cable manufacturers.

In any case the termination should not fail under handling conditions normally seen in its installed environment. The termination process should address this issue by ensuring that the correct strain relief is adopted to prevent the specified tensile or torsional loads being applied to the fiber/ferrule structure.

The second aspect of mechanical stability is that of the fiber/ferrule structure itself. It is vital that the fiber, once terminated, remains firmly fixed within the ferrule. Failure in this area may lead to fiber protrusion which could cause damage to other connector end-faces. There are two methods of securing the fiber within the ferrule – adhesive (epoxy) or friction (crimping). The former is the only truly effective method of bonding the fiber to the connector and epoxy-polish terminations are used where

126 Fiber Optic Cabling

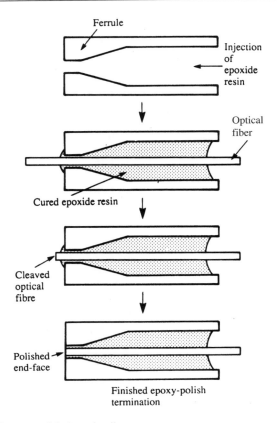

Figure 6.9 *Epoxy polish termination*

long-term performance is required (see Figure 6.9). The crimp–cleave style (frequently marketed as rapid terminations) may provide adequate fiber anchorage but these types of connectors can suffer from thermal cycling effects, such as pistoning, where the fiber pushes out of, or shrinks back into, the ferrule, according to the temperature (see Figure 6.10).

Also included in the fiber–ferrule structure is the fiber end-face. Stresses resulting from poor cleaving, adhesive curing and polishing may create surface flaws in the fiber which, if not carefully inspected, could limit the life of the termination due to cracking or chipping (which not only causes the failure of the termination but may damage mated components).

Induced optomechanical stresses (microbending)

The crimping, cleaving, glueing and polishing of an optical fiber within a fiber optic connector can create stresses (compressive, tensile and shear)

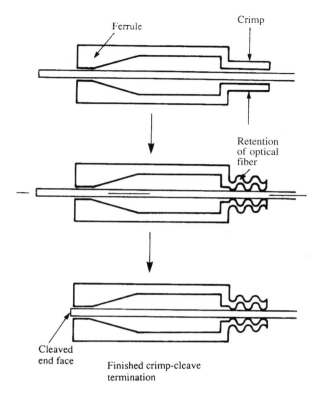

Figure 6.10 *Crimp–cleave termination*

and the stresses in turn can create losses due to microbending at the core–cladding interface.

Termination as an installation technique

The natural approach to applying connectors to cables during an installation is to terminate on site. This is a direct result of associating optical fiber with copper cable since when mains supplies, telephones and copper data communications cables are installed the contractor lays the cables and then terminates them. This practice, however, does not translate readily into optical communications and in many cases terminated cables are manufactured in a purpose-built facility and permanently jointed to the network as discussed earlier in this chapter. The justification for this approach is detailed below.

Fiber optic systems are designed and installed with the aim of carrying potentially large amounts of data over considerably greater lifetimes than is normally seen with copper systems. This places a greater emphasis upon reliability and uptime (the opposite of downtime). As the demountable connector is known to be the weakest link in the network chain it is correspondingly likely that system breakdown will occur first at a point of flexibility such as a patch panel or even at the equipment interface.

The repair via replacement of a demountable connector is not the simple task found with copper cabling connectors. The inexperienced repairer may well fail repeatedly in an attempt to rebuild the network. The option is to recall the installer or take out a repair contract on the optical network. These alternatives take time which is potentially costly in terms of the services lost to the user.

The obvious solution is to install a network design which offers simple repair tasks to the user without the need of any specific fiber optic experience. This suggests repair without retermination on site.

If a design can be produced which removes the need for repair via retermination, then it is necessary to assess the need for on site termination in the first place. In the commercial world the place to begin this assessment is the cost analysis.

The successful completion of a termination is achieved at a cost comprising the following elements:

- unit cost of components;
- unit cost of labour;
- unit cost of tooling.

In an effort to increase uptake of their connectors the manufacturers have tended to concentrate their marketing upon the perceived cost of terminating the connectors. Frequently the unit labour cost is highlighted with manufacturers offering fast-fit or rapid termination designs. In general these styles of connector achieve this apparent reduction in termination time at the expense of mechanical stability (by the removal of the adhesive bond between fiber and ferrule) or by complicating the process (thereby increasing the skill levels required by the operator to produce an acceptable result). There have even been cases where the fiber end-face surface finish has been sacrificed in an attempt to demonstrate increased throughput.

Similarly there have been occasions where the tooling cost of the termination has been the subject of perceived reduction. Unfortunately the tooling lists produced by the connector manufacturers rarely include all the items necessary for a comprehensive kit (suitable for terminating other connectors and other cables) and as a result the cost tends to be very misleading.

Finally the cost of components can only be assessed when the yield is known. The cheapest connector available could become very expensive if the eventual yield was only 10%. Components that achieve low unit purchase cost by increasing the difficulties associated with their termination are not very cost effective.

Table 6.6 Termination (field) tool kit costs

Description	Cost (multimode only)
Microscope	£400
Cleaving tool	£300
Oven	£350
Tools	£225
Total	£1275

The cost to fully equip a termination operator (exclusive of training) is of the order of £1275. This cost is detailed in Table 6.6 and includes a full range of hand tools (many of them standard copper-cabling tools) together with microscopes, fittings and specialist optical fiber tools. In general these items are not capital items and even if they can be classed as such the depreciation is rapid. On top of this one must add the cost of multimode light source and optical power meters to accurately test and measure the resulting connectorized fiber. This will be about £1500. Therefore the cost of ownership must be estimated at £2775 per annum (£12.10 per working day) minimum.

This cost is valid independent of whether the terminations take place in a factory or on site; however, the unit labour and component costs can be significantly higher on site. The reasons for this are related to volume production. It must be remembered that the replacement of termination on site by jointing of preterminated assemblies must be analysed in terms of the total cost.

X = the cost of terminating on-site
   = unit labour cost plus
     unit component cost plus
     unit tooling cost
Y = the cost of jointing preterminated assemblies
   = unit cost of preterminated assembly plus
     unit jointing cost

Termination on site can be highly time consuming, particularly if adhesive-based techniques are used (and as has already been said this technique is necessary to provide stable terminations). The quantity of successful terminations produced per day depends upon experience, the environment and the method of working; however, it is unlikely that an operator will achieve in excess of 20 terminations and more realistically the number may be 12 or less. Inexperienced operators may achieve fewer than this due to the yield implications of intermittent working. An experienced operator working 200 days per year should typically achieve 86% yield overall and produce 16 terminations of assessed quality per day. An intermittently operated team working 50 days per year may achieve 67% yield overall and produce only eight terminations per day.

It is understandable if the reader regarded the number of terminations completed per day as being rather low. A termination manufactured in a factory environment will take approximately 7 minutes to complete (prior to inspection and testing) and it might be thought that a typical value for on-site work might be similar. Unfortunately on-site working including pretesting of installed cables and equipment set-up may increase the time per termination to approximately 15 minutes or even longer. Increases in output can be achieved only at the expense of quality (by using crimp–cleave technology or by accepting inferior inspection standards, both of which risk long-term performance) or by using preterminated or prestubbed connectors The latter is a factory-finished connector with a mechanical optical splice behind it. The quality is therefore very high for the termination and all the operator has to do is accurately cleave the fiber and put it into the back of the connector and crimp it in place. Unfortunately the material costs are generally three times higher than for standard epoxy-cure and polish connectors, but for the infrequent user this may still be the most cost-effective route.

The analysis above is based upon standard multimode components and shows that there is a marginal advantage to jointing preterminated assemblies on-site rather than attempting to produce terminations of assessed quality on-site.

Summary

This chapter has reviewed the options open to the installer for the jointing of optical fiber. The division of jointing techniques into permanent and demountable allowed an assessment of the advantages and disadvantages of each type.

For permanent jointing, fusion splicing, whilst offering the best performance levels, can be expensive for the intermittent installer; however, mechanical splices tend to be too expensive for regular use.

The demountable connector market has undergone considerable change and growth. Demountable connectors are relatively complex to apply to cable and involve much more skill than is immediately apparent.

The decision process for connectors will be along the lines of:

- What kind of connector? e.g. ST, SC, MT-RJ etc.
- Where will the connector be presented? The connector used at the wall outlet may not be the same selected for the main patch panel or campus distributor.
- Single mode connectors also have the choice of PC, SPC, UPC, APC finish.
- Direct termination on site:
  - epoxy, heat cure and polish, 'pot and polish';
  - 'hot melt';
  - anaerobic/cold cure;
  - 'crimp and cleave'.
- Prestubbed connectors with in-built mechanical splice.
- Fusion splicing or mechanical splicing of factory-made tail cables.

Tail cables are short lengths of fiber with a factory fitted connector on one end. Most users apply these when terminating single mode fiber. Single mode connectors can be put on on-site, but the yield is reported to be very low, i.e. the failure rate is very high.

References

1 Cotruta, T., Optical interoperability, why we need it, how we get it. *Lightwave*, October 2000.
2 Telcordia, GR-326-CORE, Generic requirements for single mode optical connectors and jumper assemblies, Issue 3, September 1999.

7 Fiber optic cables

Introduction

Having discussed the theory, design and manufacture of optical fiber in Chapters 2 and 3 it is logical at this point to move on to the issue of fiber optic cable. The cabling of primary coated optical fiber is an important process for two reasons; first, it protects the fiber during installation and operation and, second, it is responsible for the environmental performance of the contained fiber. This implies that the methods used to cable optical fiber elements can, if not controlled, severely impact both the initial and long-term performance of an installed network. This chapter discusses the basic variants of cable constructions, their suitability in given environments and finally details the possible problems induced by incorrect cabling procedures.

Basic cabling elements

There is a bewildering variety of fiber optic cable designs available, each manufacturer often having a particular speciality. However, all cables contain fiber in one of three basic elemental forms.

Primary coated optical fiber (PCOF)

Primary coated optical fiber is the end product of the optical fiber manufacturing process (see Figure 7.1). It is remarkably strong and stable under tensile stress but will fracture if subjected to the excessive amounts of bending and twisting associated with installation. Nevertheless it is widely used as a basic element within cable constructions which are themselves capable of withstanding the rigours of installation.

As the PCOF undergoes no further processing before being cabled it is by far the lowest-cost option within a cabling construction but the

Fiber optic cables   133

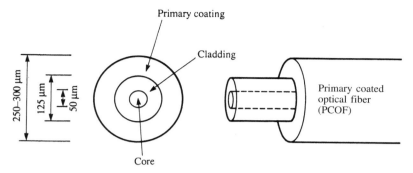

Figure 7.1 *Primary coated optical fibre*

resulting cables can be awkward to terminate with large numbers of 250 micron fibers arriving at one place. External grade/telecommunications/high fiber count cables tend to use primary coated fibers protected within a larger tube, called a loose tube construction, but premises cabling often uses secondary coated (tight buffered) fibers for their ease of termination.

Secondary coated optical fiber (SCOF)

Secondary coated optical fiber, often referred to as tight-buffered fiber, features an additional layer of plastic extruded on top of the PCOF (see Figure 7.2). The resulting element is typically 900 μm in diameter and can be incorporated within much more flexible constructions as are required within buildings or for use as patch or jumper cables.

The application of the secondary coating necessitates a further production stage so cables containing SCOF tend to be more expensive.

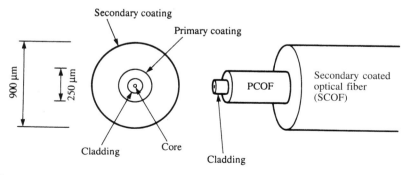

Figure 7.2 *Secondary coated (tight-buffered) optical fibre*

Single ruggedized optical fiber cable (SROFC)

Optical connectors cannot be applied to the PCOF or SCOF with any real level of tensile strength. As a result some form of strength member must be built into the cable construction. At the individual element level the most basic structure containing an effective strength member is the single ruggedized optical fiber cable.

In this construction an SCOF is wrapped with a yarn-based strength member which is subsequently oversheathed with a plastic extruded material. The yarn may be aramid (Kevlar®) or glass fiber (see Figure 7.3). As a result the SROFC element is the most costly format in which to provide an individual optical fiber.

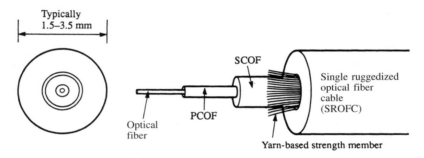

Figure 7.3 *Single ruggedized optical fiber cable*

All the above basic cabling elements can be used either singly or in multiples within a larger cable structure. The final choice of cable design is highly dependent upon the application and the installed environment, and, to a large degree, upon how the cable will be installed.

Cabling requirements and designs

Fiber optic cables are required to be installed and operated in as many different environments as are copper cables. It is not sufficient to merely ask for an optical fiber cable; its final resting place may be buried underground, laid in a water-filled duct, strung aerially between buildings (and subject to lightning strike) or alternatively neatly tied to cable tray running through tortuous routes within buildings. The cable design chosen for a particular installation must take account of these conditions and it should be pointed out that a given installation may actually use more than one design of cable.

Fiber optic cable design definitions

To assist in categorizing the large number of different designs available it is necessary to provide some definitions relating the application of a cable to its design.

Fixed cable

The strict definition of a fixed cable is one which, once installed, cannot be easily replaced. So rather than defining a design of cable the term defines its application. However, most fixed cables have a common format and contain one or more optical fibers which do not have individual strength members. Such cables normally have a structural member which provides the desired degree of protection to all optical fiber elements during installation and operation.

Cables of this design cannot be properly terminated without further protection for the individual optical fibers. The cables must be terminated in an enclosure fitted with suitable strain relief glands which are connected to the structural strength member within the cable.

The enclosures, defined in this book as termination enclosures, are themselves fixed to ensure that no damage can be done to the optical fibers by accidental movement of the cable or enclosures. As a result the cables between the terminating enclosures are termed fixed cables. Fixed cables are the most varied in design since they tend to be used in the widest range of installed environments.

For cables running between buildings, the fiber count (the number of individual optical fiber elements) can vary from one to well over a hundred and, to reduce both cost and overall diameter, the cables tend to contain PCOF lying in a loose format within the cable structure (see Figure 7.4).

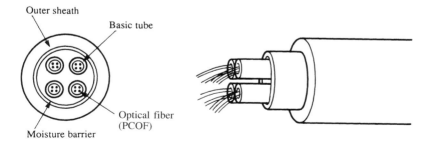

Figure 7.4 *Loose tube cable construction*

136 *Fiber Optic Cabling*

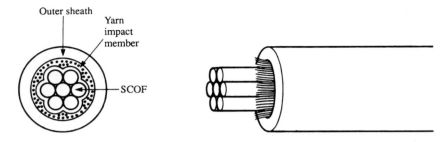

Figure 7.5 *Tight jacket (buffered) cable construction*

The multiple-loose tube versions of these cables have the disadvantage of being rather rigid and tend to be non-ideal for intra-building installation where tight radii are often encountered. Fixed cable designs are available containing SCOF laid in a tighter structure (see Figure 7.5) which allow more flexible cable routeing to be adopted. Single loose tube, or unitube, designs do allow for a flexible, smaller PCOF design.

In a specific installation the primary route may comprise more than one fixed cable design (perhaps a mixture of SCOF and PCOF formats) with the cables being permanently jointed or demountably connected at termination enclosures.

Fixed cable looms and deployable cables

For specific, well-defined, but short-range installations (for example, airframes, vehicles and smaller or temporary premises cabling installations) the fixed cables may consist of a cable loom containing SCOF elements directly terminated with either single or multi-element connector components. These are manufactured prior to installation and may not be straightforward to replace. For this reason dual or triple levels of redundancy are designed into the installation. If replacement is easy then these looms are more akin to jumper or patch cables (see below).

In special cases field deployable cables can be produced containing one or more SCOF elements in a tight construction which may be directly terminated with specially designed multi-way connectors. Cable strain relief is achieved within the overall connector body. This type of construction does not strictly represent an installation at all since the flexibility of the cable assembly lies in its ability to be deployed and re-reeled at will.

Patch or jumper cable

The fixed cable is normally a multi-element cable which cannot be terminated directly without the protection of a termination enclosure. The termination enclosure may be connected to other termination enclosures or to transmission equipment by either patch or jumper cables respectively. Occasionally two pieces of transmission equipment may be directly connected using jumper cables. A schematic representation of a network comprising all these features is shown in Figure 7.6.

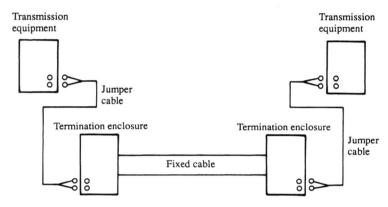

Figure 7.6 *Cabling schematic*

Patch and jumper cables differ only in their application. The design is the same for each. They are a means of achieving cost-effective, reliable and easily replaceable connection between termination enclosures and equipment. The cables normally comprise SCOF in single or duplex format (although the example of looms highlighted above may include very many individual SCOF elements) and are directly terminated in the factory. The necessity for effective strain relief throughout the assembly is paramount since these cables will be subject to handling (usually non-expert).

As fiber optic transmission is typically unidirectional then duplex transmission requires the allocation of two optical fiber elements within a cable. It is therefore not unreasonable for users to request duplex jumper or patch cables.

The above broad definitions of cable types and their usage is useful not only as a method of categorizing any particular offering but also as a means of splitting up a given installation into fixed cable, jumper or patch cable sections. This enables a strategic view to be taken towards the installation, the procurement of the components to be used and the repair and maintenance philosophy to be adopted following installation. The following section reviews the detailed requirements for cabling in specific applications.

Inter-building (external) cables

When optical fiber was first adopted within the telecommunications industry the cabling requirements were largely external. Many kilometres of directly buried or duct laid cables were jointed together and entered buildings only at transmission or repeater stations. The cabling designs became very standardized and were purchased in huge volumes.

However, in the wider area of application commonly termed the data communications market some of these designs were not necessary or cost effective, and as a result many new custom-built configurations have been produced. Obviously this text cannot describe in detail every possible format and this section (and that covering intra-building cables) can only highlight the key issues which must be addressed by the cable manufacturer and installer.

There are a number of ways in which a cable can be installed between buildings. To summarize, these are direct burial, laying in ducts or existing cabling run (tray or pipework) or aerial connection (by catenary/messenger wire structure or self-supporting). Each environment throws up some specific requirements which can all be met with existing technology.

All external fixed cable formats are designed to undergo physical hardship during installation. It is important that the optical fibers contained within the structure do not suffer from the rough treatment received by the external surfaces of the cable or from the tensile loads applied to the cable strength member during installation. In most cases therefore the optical fiber lies loose within the structure of the cable, and as a consequence it can be provided in PCOF form. As has already been said, the cost of PCOF is relatively low and for external fixed cables it is fair to say 'fiber is cheap, cabling is expensive'.

From the cost viewpoint, the installation of the cables, including any necessary civil works and provisioning, will be significant, and it makes good economic sense to minimize the probability of having to install further cables at a later date. It is sensible to include spare fiber elements within external fixed cables wherever possible, even if they are not commissioned during the initial installation. Alternatively, empty plastic tubes can be installed and the optical fibers blown in at a later date when they are required. This technique is generically called blown fiber, or sometimes ABF, air-blown fiber, and requires specially coated fibers and ducts for the fiber blowing to be effective. The blown-fiber ducts are generally 5 millimetres in diameter. One method blows in a group of fibers encapsulated into one unit, whereas another style, Blolite®, blows in up to eight individual fibers into each tube. The latter method is best for negotiating tight corners (down to 25 mm) as may be encountered within premises cabling. The former is better suited to longer, straighter runs as will be more often seen in inter-building cable routes. The individual blown fibers are akin to tight-buffered, secondary coated units, but are oversheathed to only 500 μm (with a low friction, anti-static material) compared to 900 μm diameter of conventional SCOF units.

In the external fixed cables the PCOF elements will be contained within tubes or extruded cavities which surround a central strength member, known as slotted core. Further layers of cable construction surround these tubes and it is these layers which provide the environmental and installation protection for the optical fiber.

Virtually all external fixed cables will be subject to attack by moisture. The moisture can enter the cable in two ways: first damage to the sheath layers during installation could allow penetration of moisture into the cavities, and second moisture could enter from unprotected ends of the cable (at underground jointing enclosures, for example). To prevent the first from occurring, cables can be manufactured with moisture barriers which surround the fiber cavities. Alternatively the cable can be provided with damage-resistant sheaths or moisture-retention layers which achieve high levels of protection (at additional cost). The second mechanism, moisture travelling along the optical fiber, can be prevented by the inclusion of a gel material within the cavities. Whilst this is undeniably effective it also creates problems at the installation stage and is normally adopted only where totally necessary. A more modern technique for premises and campus cabling is to incorporate water-swellable tapes and threads within the spaces within the cable (the 'interstices') which swell up if in contact with water and hence seal off the longitudinal path for water ingress.

The choice of moisture-protection technique depends upon the installed environment. Direct burial cables and cables to be laid in water-filled ducts will be subject to almost continuous attack from moisture at their surface and must be provided with an effective moisture barrier. This normally takes the form of a polyethylene and aluminium laminate which is wrapped around the fiber cavities as the cable is made but prior to final sheathing. Such a cable would also exhibit a central strength member of braided steel wire (for example) which would allow the cable to withstand the tensile loads seen during installation. These cables cannot be termed metal-free and are therefore capable of creating problems during lightning strikes. Also earthing of the metallic elements is a continuing cause for concern and must be accommodated where the cable enters a building or equipment room.

Where a metal-free cable is required, for example in aerial installations where the cable is tied to existing catenary structures, then the steel strength member must be replaced with an insulating material such as glass-reinforced plastic or aramid yarn. Similarly the moisture barrier must be assessed in terms of its ability to conduct the large electrical content of a lightning discharge. A non-metallic moisture barrier is necessary and can be provided in a number of ways but one very effective method is to produce a sandwich of yarn-based materials between the outer sheath and an internal sheath. Any damage to the external sheath allows moisture to pass on to the yarn sandwich (which will also act as an impact-absorbing layer) but not to pass any further. A totally metal-free cable of this design is suitable for aerial applications and, where the tensile strength allows, such cables could be used universally in all applications.

Direct burial cables normally feature some type of armouring using steel wire or corrugated steel tape. The armouring is present to prevent damage during installation but it is equally important in preventing damage due to external influence following installation.

Moving away from the moisture and structural members within the cable the sheathing materials are worthy of discussion. The vast majority of external fixed cables are sheathed in polyethylene or polypropylene. These materials feature hard, low-friction surfaces that are ideal for installation environments where abrasion resistance is important. In addition they exhibit good levels of moisture resistance. The newer low fire hazard (LFH) materials used in internal fixed cables are spreading to the external environment. These new materials when combined with polyethylene bases provide the best of both worlds allowing the external cables to be used over extended distances within buildings (thereby removing the need to have external–internal joints).

The external fixed cable is a complex combination of materials of which the optical fiber is almost the least important (since it is unaffected by the surrounding construction). The features of the cable are largely determined by the application and installed environment. The above discussion is summarized in Table 7.1.

Table 7.1 *Design requirements of external fixed cables*

Application	Design features
Direct burial	• Steel wire or corrugated steel tape armoured • Moisture barrier • Central strength member, probably steel • Gel-filled, loose tube
Cable duct	• Central strength member, steel or GRP • Moisture barrier • Gel-filled, loose tube
Aerial	• Loose tube • Central strength member • Non-metallic construction

Intra-building (internal) cables

The methods used to install external fixed cables determine the design of the cables where the optical fibers, normally in a PCOF format, lie loose in cavities or tubes. The tubes are themselves held within a construction which may include metallic or non-metallic strength members, moisture barriers and various layers of sheathing materials. Needless to say the resulting cables are not very flexible, with diameters above 10 mm. Estimates of minimum bending radius normally lie in the region of 12 × cable diameter and as a result the cable cannot be bent to a radius less than 150 mm (6 in) and this limitation can be restrictive within buildings.

Internal fixed cables are designed to overcome this limitation. They achieve flexibility at the cost of mechanical strength but are a vital component in the installation of optical fiber networks.

Internal fixed cables are available in a variety of optical fiber formats and these are briefly discussed below.

Breakout internal fixed cables

These are the most rugged of the options and they contain a number of individually cabled tight-buffered elements in an overall sheath. As the elements themselves have diameters of between 2.4 and 3.4 mm the resulting cables can be quite large. They are most frequently used where the ends of the cable are to be directly terminated.

Compact internal fixed cables

These cables combine flexibility with compact design. A number of SCOF elements are wrapped either around a central former or within a yarn-based layer. These cables are flexible (minimum bend radii of 50 mm) and are lightweight. The amount and design of yarn-based wrap does vary from a token presence (acting as an impact resistance layer) up to a full and very tightly bound construction capable of withstanding significant tensile loads (applied to the wrap). However, as the cost of the cable increases dramatically with the yarn content the latter designs are normally seen only in military applications such as field deployable communications.

PCOF internal fixed cables

The use of PCOF elements within internal fixed cables is limited since the difficulty in providing satisfactory impact resistance limits the achievable flexibility of the finished cable. Loose tube cable constructions merit additional care at the installation stage and tend to work against the underlying aim of producing cables that are no more difficult to install than copper communications cables.

Internal cable sheath materials

The materials used to sheath the internal fixed cables have undergone significant changes over the last few years. At one time the standard material was polyvinyl chloride (PVC); however, there has been a definite trend towards a range of materials broadly described as low fire hazard or low smoke and zero halogen.

There are many standards around dealing with the effects of fire on cables within a building, but until the Construction Products Directive comes into force in the European Union, very few of them are mandatory. However, many large users now specify zero halogen, low flammability cables as an extra safeguard for their own premises, information technology equipment and personnel. The appropriate IEC standards for cables and fire performance are:

- IEC 60332-1 Flammability test on a single burning wire or cable.
- IEC 60332-3-parts 21, 22, 23, 24 and 25
  Flammability test on a bunch of wires or cables.
- IEC 60754 Halogen and acidic gas evolution from burning cables.
- IEC 61034 Smoke density of burning cables.

IEC 60332-1 is seen as the base level of fire performance for intra-building cables. Note that halogenated material such as PVC can still meet tough fire tests, but cannot of course meet the zero halogen tests. CENELEC will take the IEC tests, and add some of their own, for European requirements:

- EN 50265-2-1 IEC 60332-1 flammability, single cable.
- EN 50266-2-4 IEC 60332-3 flammability of a bunch of cables.
- EN 50368 IEC 61034 smoke evolution.
- EN 50267 IEC 60754 acidity and conductivity.
- EN 50289-4-11 Flame propagation, heat release, time to ignition, flaming droplets.

The 2000 CPD proposals for cables within the European Union are shown in Table 7.2.

In the United States, fire regulations for cables to be used within buildings have been around for much longer. Most cables are specified as riser grade or plenum grade. Plenum means an architectural space where environmental air is forced to move. This usually means the ceiling void that is also the return path for air in air-conditioned buildings. The UL 910 test for plenum cables is quite severe, and so far only halogenated

Table 7.2 *2000 Construction products directive proposals*

Fire situation	Euroclass	Class of product
Fully developed fire in a room	A	No contribution to the fire
	B	Very limited contribution to a fire
Single burning item in a room	C	Limited contribution to a fire
	D	Acceptable contribution to a fire
Small fire effect	E	Acceptable reaction to a fire
	F	No requirement

Table 7.3 *American internal optical cable marking scheme*

Cable title	Marking	Test method
Conductive optical fiber cable	OFC	General purpose UL 1581
Non-conductive optical fiber cable	OFN	General purpose UL 1581
Conductive riser	OFCR	Riser UL 1666
Non-conductive riser	OFNR	Riser UL 1666
Conductive plenum	OFCP	Plenum UL 910
Non-conductive plenum	OFNP	Plenum UL 910

materials such as PTFE, polytetraflouroethylene (better known by its brand name of Teflon®) is capable of meeting this test. Table 7.3 gives the American optical cable ratings.

Fiber optic cables and optomechanical stresses

The preceding sections in this chapter have discussed the basic cabling components of primary and secondary coated fibers and their incorporation into the larger cables used both as inter- and intra-building cables. The optical fiber has been treated as a mechanical component within the larger structure and no optical performance issues have been addressed. However, the cabling of optical fiber does influence the overall performance of the fiber both at bulk and localized measurement stages.

Cable specifications

Optical fiber manufactured as PCOF is measured against defined attenuation and bandwidth specifications. In addition the physical parameters such as core diameter, cladding diameter, core concentricity and numerical aperture are checked against the manufacturing specification. The optical fiber is then purchased by, and shipped to, the cable manufacturer and it will be processed into cable in one of the many formats described earlier in this chapter. Depending upon the final construction of the cable the optical fiber itself may or may not have received further direct processing (from PCOF to SCOF or SROFC, for example) and this processing may or may not have influenced the bulk optical performance of the PCOF as purchased.

Normally a loose PCOF element within an external fixed cable will not exhibit any significant change in performance since the optical fiber is under little or no stress within the construction.

The application of a 900 µm secondary coating to a PCOF may modify its performance and the cabling of these elements in a tight construction can markedly affect the final attenuation measured. In this way a change in attenuation is a measure of stress applied to the optical fiber.

As a consequence when purchasing cable it should be clearly stated that the attenuation measured is of the cabled optical fiber and not the original PCOF.

Cable and fiber specifications are now covered under the following generic standards:

CENELEC
- Optical fiber   EN 188 000
- Optical cable   EN 187 000

IEC
- Optical fiber   IEC 60793
- Optical cable   IEC 60794

To expand IEC 60794 in more detail:

- IEC 60794-1-2 Testing
- IEC 60794-2   Internal cables
    IEC 60794-2-10 Simplex and duplex cables
    IEC 60794-2-20 Multi-fiber cables
    IEC 60794-2-30 Ribbon cords
- IEC 60794-3   External cables
    IEC 60794-3-10 Duct or buried cables
    IEC 60794-3-20 Aerial cables
    IEC 60794-3-30 Underwater cables
- IEC 60794-4   Cables along electrical overhead lines

Some American optical cable standards are:

- ICEA S-87-640 Optical fiber outside plant cable
- ICEA S-83-596 Optical fiber indoor/outdoor cable

The specified optical cable performances from ISO 11801 2nd edition, EN 50173 2nd edition, and TIA/EIA 568B are shown in Table 7.4.

Table 7.4 *Cable performance requirements from premises cabling standards*

	Maximum cable attenuation, dB/km ISO 11801 2nd edition			
	Multimode		Single mode	
Wavelength	850 nm	1300 nm	1310 nm	1550 nm
Attenuation	3.5	1.5	1.0	1.0

Connector-based losses

The design of cable can significantly modify the losses experienced when the cable is terminated. The losses within a mated connector pair are generated by the tolerances of the connector and the optical fiber but they can be radically altered by stress applied to the fiber within the rear of the connector. These are microbending losses as discussed in the previous chapter.

Microbending losses are most frequently seen in terminations involving sprung ferrules where the mating action of the connectors tends to force the fiber back into the cable construction. If the design of the cable is such that the optical fiber is tightly packed, then the fiber tends to become compressed within the connector. This results in losses that increase rapidly as the connector is mated with the effect being exaggerated in the second and third windows (1300 nm and 1550 nm).

However, it is not only cable construction that can affect termination performance. Badly produced PCOF, where the application of the primary coating has been faulty, or poor monitoring of secondary coating procedures can lead to the manufacture of optical cable that cannot be terminated in any sensible fashion since externally applied stress, such as that from a light crimp (as used to connect the back end of the connector to the cable construction), is seen to create large microbending losses. These faults may not be detected at the time of cable manufacture and vigilance is vital if the cable is not to be passed through to be terminated.

It is fair to say that the inexperienced user will be no match for the cable manufacturer at the technical level and if there is any doubt with regard to the usability of a particular cable it is important to obtain experienced assistance before wasting time and money in attempting to use that cable.

It is important therefore to ensure that the cable is always compatible with the connector to be used.

Fiber mobility and induced stress

At first glance it is tempting to assume that an optical fiber cable is a stable component without any particularly unwelcome characteristics. Unfortunately this is not always true and the further installation (by direct termination or by the use of termination enclosures) must reflect this instability or problems will be experienced which may not be resolved without significant rework.

The use of loose construction cables where single or multiple PCOF elements are contained within a sheath have particular problems which must be addressed at the earliest stage of installation. As was mentioned earlier in this chapter the most basic construction consists of a tube, perhaps 5 mm in diameter, in which a small number (normally between one and eight) of PCOF elements are laid, free to move within the tube.

Professional installation of such cables and their larger and more complex counterparts will always use a termination enclosure. First this ensures that the necessary strain relief is given to the cable, but second it allows excess fiber to be coiled within the termination enclosures, thereby removing any concerns with regard to fiber mobility within the cable. However, in an effort to cut installation costs these basic cables are sometimes directly terminated prior to installation with the cable on its reel. The problems occur when the cable is subsequently unreeled.

When the cable is on a reel the tube has a fixed length and the optical fiber assumes the most stress-free path, i.e. the shortest path (which is shorter than the tube). If the cable is then directly terminated and the connectors attached to the tube and then unreeled the fiber finds itself shorter than the tube and as a result breaks occur. Unfortunately the breaks do not always occur at the terminations since the adhesive bond may actually be stronger than the optical fiber. As a result the break is difficult to locate and cable replacement has to take place. The solution is to ensure that there is excess fiber within the cable. Unfortunately it is difficult to assess the amount of excess fiber needed and sometimes it is rather difficult to push the fiber back into the cable (thereby introducing the desired excess) at the time of termination.

The other aspect of fiber mobility is the movement of tight constructions under conditions of high tensile load. When repeated field deployment of a cable is necessary (military communications, outside broadcast etc.) it is normal for the cable to have to withstand tensile loads well beyond the limits of the optical fiber itself. This is achieved by the introduction of a variety of strength members including aramid yarns (to maintain flexibility and impact resistance). The cables are frequently terminated directly in multiway connectors which provide strain relief and protection for the optical fiber elements. The tensile loading of the cable (normally between the connectors at either end) is intended to be absorbed by the strength member and the overall extension of the strength member is designed to be less than the acceptable strain for the optical fiber. In this way the fiber cable can withstand significant loading (4000 newtons having been achieved on a 5 mm diameter cable containing up to four SCOF elements). However, the mobility of the sheath and the strength member in relation to the optical fiber is a complex issue and in some designs a variation in loading profile can dramatically affect the maximum allowable load. Equally important is the nature of the termination technique and the style of connector applied to the cable ends. Unterminated cables tend to exhibit much greater breaking loads than terminated assemblies. Also the position of the breaks may radically alter once the assemblies are terminated. As a result it is vital to perform detailed testing upon such high-specification cables not only as cables but also as cable assemblies.

Air-blown fiber should, in theory, be stress free once installed. This is because the tube is installed without the fiber and so the fiber doesn't suffer the usual stresses of installation. Once the fiber is blown in and the air supply turned off, the fiber will relax and settle into the position of least stress.

User-friendly cable designs

The cable manufacturer is in business to produce cable. It is not their responsibility to guarantee its ability to be installed or jointed or terminated with the connectors currently available.

That being said the manufacturer that supplies user-friendly cable designs tends to be warmly welcomed into the market.

User-friendly cable exhibits the following features:

- easily stripped sheath materials;
- uniquely identified fibers (by alphanumeric addressing or colours);
- easily stripped secondary and primary coatings.

The inclusion of these features within cable designs, whether they are fixed cables (external or internal) or jumper cables, makes the cables much easier and therefore less time consuming to install, joint or terminate.

The economics of optical fiber cable design

The wide range of cable designs available complicates the business of choosing the right cable for the application. As usual in all practical things there is rarely only one solution to a given requirement. Indeed, as discussed in Chapter 10, it may be that the ideal solution may not be achievable in the time scales of the project and a second option has to be adopted. In this way the cost of cable can actually become of secondary importance, availability taking pride of place.

Nevertheless an appreciation of the basic economics of optical fiber cable design is desirable, particularly at the custom-design level.

This section reviews the cost structures within an optical fiber cable and also compares the unit cable cost with the overall installation costs.

Optical fiber cost structure

As the direct result of a highly efficient volume production process, primary coated optical fiber represents the lowest-cost optical fiber structure. The cost of the optical fiber has been discussed in earlier chapters but is shown diagrammatically in Figure 7.7. It shows that single mode 8/125 μm geometries are the cheapest to produce. The 50/125 μm and 62.5/125 μm fibers show increasing costs leading to the relatively expensive large core diameter, high NA fibers such as the 200/280 μm designs.

For a given fiber the addition of a secondary coating (taking the optical fiber to perhaps 900 μm in diameter) is a separate process which must be undertaken prior to final cabling. This represents a fixed cost which must be added to the basic PCOF cost.

The cabling of the SCOF element within a single ruggedized optical fiber cable (SROFC) represents a further fixed cost. This fixed cost may vary slightly dependent upon the sheath material and the density of the yarn-based strength member included in the design.

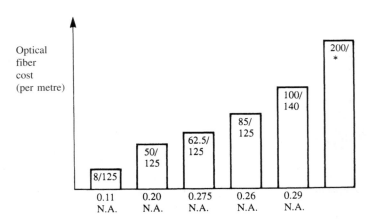

Figure 7.7 *Optical fiber cost versus geometry*

The cheapest cables comprise PCOF elements and the most expensive are manufactured using SROFC units. There are two other factors which must be taken into account in order to establish the most cost-effective design for a particular application. The first is the cable structure and the second is the method of installation.

Cable cost structure

As has been stated throughout this chapter the cable construction must be capable of providing protection to the optical fiber both during installation and during the extended operational life predicted for the cable.

Based upon the optical fiber cost structure it would appear sensible to use PCOF elements in all cases. Unfortunately the cable constructions necessary to provide protection to PCOF elements tend to limit the flexibility of that cable. Also the need to terminate the optical fibers directly may restrict the use of PCOF elements and favour SROFC instead. The application may therefore modify the apparent cost benefits produced by the use of PCOF within a cable.

Fixed external cables are frequently used in considerable quantities particularly in the multi-building network type environment. Their construction is normally loose (i.e. the fibers lie within the construction under little applied stress) which suggests the use of tubes or a former within the overall cable construction. To standardize on the designs a number of tubes or formers will be included independently of the number of optical fibers included within the cable. This represents a fixed cost structure. The existence of a central strength member, laminate moisture barrier and the inclusion of a gel-fill similarly can be thought of as fixed costs.

The fixed cost element of a fixed external cable can totally mask the low optical fiber cost generated by using PCOF elements. Figure 7.8 indicates the cost curves of such a cable design.

Fixed internal cables have little need for moisture performance and are normally required to be flexible and, at the same time, impact resistant. This limits the use of loose PCOF constructions and SCOF elements are often used, wrapped with a strength/impact layer and oversheathed. The fixed costs in this type of construction are lower than for the external PCOF designs but the optical fiber costs are higher. The cable cost versus fiber count equation is therefore more linear as is shown in Figure 7.8.

The most linear cable cost equations are those of cables containing SROFC elements. Each element contains its own strength–impact resistant member and its own oversheath. The construction of such cables is normally completed by the application of a simple oversheath. However, the basic elements are far more expensive than those of SCOF or PCOF and are only really viable when related to the overall cost of installation.

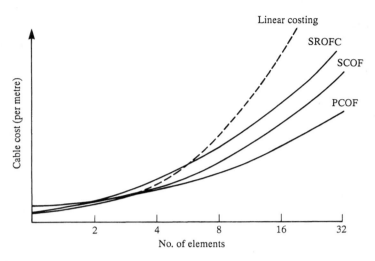

Figure 7.8 *Cable cost versus element count*

Installation cost structure

Based upon the information summarized in Figure 7.8 it would appear that the lowest-cost design will comprise PCOF in a design having the lowest fixed construction cost. Such cables exist and are normally seen to comprise a single tube into which a number of PCOF elements are laid. No moisture barriers are included and strength members are absent or are minimal.

Naturally it is tempting to choose such a design; however, the costs of installation must be considered and balanced against the apparent savings. Obviously this cable has a number of disadvantages, the first of which is the absence of a moisture barrier (which prevents its use as an external fixed cable), the second its susceptibility to bending and kinking during installation. The latter can be addressed by either taking more care with the installation, which will increase the overall cost of the cable, or by replacing those sections damaged (which has the same effect).

It can be seen therefore that the most cost-effective external fixed cable must include PCOF with full environmental protection. It is also seen that it is sensible to include as many optical fiber elements as is feasible since the additional costs are not linear. This argument is developed further in Chapter 9.

The same cable could be used for internal fixed cables; however, the issues of sheath toxicity and flexibility tend to limit its use. Looking towards both SCOF and SROFC designs it is clear that the SCOF formats will be cheaper. However, SROFC formats may be cheaper to install since it is possible to preterminate one end (or even terminate *in situ*) without the use of a termination enclosure. Therefore the overall cost of installation can be a deciding factor in the choice of an internal fixed cable, which can override the basic cost structures of the cables themselves.

Summary

This chapter has defined the various types of optical fiber cable both by application and design. In addition the issue of cost structures has been addressed which may assist the potential user and installer alike in the choice of cable designs to meet the specific requirements of an installation.

Cable design and choice is reviewed again in Chapter 10 when the commercial practices and compromises found in realistic installations are discussed.

8 Optical fiber highways

Introduction

The preceding chapters of this book have dealt with the theory and practice of production of optical fiber and cable, optical connectors and their application (or termination) to fiber optic cable and the methods of jointing one piece of optical fiber to another. This information and the skills developed from it are sufficient to install the vast majority of optical fiber highways. This chapter reviews the nature of such a highway and discusses the common aspects of all installations.

Optical fiber installations: definitions

All optical fiber communication assumes a common format. Information is transmitted from one location to another by the conversion of an electrical signal to an optical signal, the transmission of that optical signal along a length of optical fiber and its reconversion to an electrical signal.

This communication may take place between two or more locations, creating the concept of a network of communicating centres or nodes. These nodes may be close together, as in an aircraft, or many kilometres apart as in a telecommunications network. It is useful therefore to produce some definitions to allow standardization of terms within this text.

Optical fiber link

The optical fiber span is the most basic cabling component between two points which can be considered to be individually accessible.

Using this definition a jumper cable or a patch cable is an optical fiber span. Both are individually accessible from both ends via the demountable connectors attached to the simplex or duplex cable formats. Similarly a long multi-element fiber optic cable featuring many fusion splice joints and terminated in patch panels is also an optical fiber span since access can only be gained at the connections to these patch panels. This can be readily extended to include a path which comprises a number of different types of fixed cable (external and internal) jointed together and accessed by connection to spliced-on connector tails.

Optical fiber highway

The optical fiber highway is an open configuration of fixed cabling passing between a number of nodes. The fixed cabling between each node is an optical fiber highway in its own right and may consist of more than one optical fiber span. The term optical fiber highway applies equally to the totality of all fixed cables within a given installation.

The concept of the optical fiber highway is intended to reflect the open nature of the cabling and is not related to the purpose to which the optical fiber within the cabling is to be put. To this end the concept of the optical fiber highway does not include jumper or patch cables which are used to configure the highway to provide the various services required. That is to say the optical fiber highway may distribute a range of services from node to node or between distant nodes and can therefore be regarded as an open infrastructure which can be configured to meet a changing set of requirements. To some extent this reflects the inherently large bandwidths of the optical technology that allows continually upgraded services to be run on the same cabling infrastructure.

Optical fiber network

The optical fiber network is a term which describes the usage of the optical fiber highway at a moment in time. A given optical fiber highway can be simultaneously operating a token ring network, an Ethernet network, some point-to-point services such as video surveillance or CADCAM and perhaps some basic telemetry signalling. The optical fiber highway is therefore configured by the appropriate selection of patching facilities to provide this network of communications. The optical fiber network can be thought of as an overlay on top of the optical fiber highway.

The optical fiber network can connect both active and passive nodes. An active node is a location that provides optical input and/or receives optical output from the highway whereas a passive node is merely visited. In this way a given node may be passive for one optical fiber network overlay but active for another.

Summarized, the optical fiber network is the service configuration of the optical fiber highway, which merely acts as the service provider.

Optical fiber system

The optical fiber system includes the transmission equipment chosen to initiate the services to be operated on the optical fiber highway. The trend away from turnkey communications installations on optical fiber as the market has matured has led to a diminishing use of the term. Since any one installation may operate many different types of transmission equipment it is difficult to refer to a system. It is more common in these cases to talk of an optical fiber highway operating multiple systems.

The above definitions are shown in diagrammatic form in Figure 8.1.

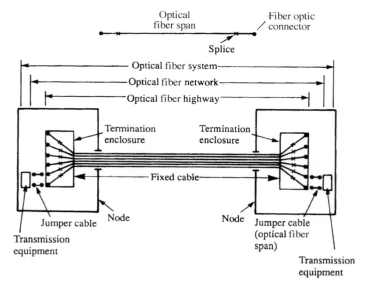

Figure 8.1 *Optical fiber highway definitions*

The optical fiber highway

With the exception of the most basic trial or prototype systems the installation of an optical fiber highway should not be treated as a trivial matter. A large number of organizations have adopted this approach and have regretted their mistake. A lack of understanding of optical fiber has certainly contributed to their downfall; however, the more deep-rooted problem has been one of not understanding the nature of communication cabling as a capital purchase which is expected to provide a return on investment.

Optical fiber, as has been seen in the early chapters of this book, has a major role to play as part of a wider communications structure. Its high bandwidth and low attenuation suggest its installation on major routes within that structure. These major routes will in general be easy to identify and are likely to be permanent.

In the telecommunications area these permanent routes are clearly defined, being trunk routes and the linkage to the local exchanges. In the data communications environment these major routes are mimicked by the inter-building structures on large sites (e.g. hospitals and universities) or inter-floor structures within single buildings. Both of these are permanent from the viewpoint that neither the buildings nor the floors are likely to move in the foreseeable future.

Moving away from these two examples it is worthwhile to identify where the optical fiber highway concept is relevant within the military market. As has already been stated the application of optical fiber to military communications comes in many forms. The introduction of optical fiber into surface ships, submarines, fighting vehicles and aircraft is subtly different from their application for land-based communication. The latter tends to follow the well-trodden path of telecommunications and commercial data communications and may adopt fixed or field-deployable cabling strategies. The other applications involve relatively short-range, high-connectivity highways which can nevertheless be regarded as permanent and major routes since their replacement will not be straightforward and changes in their routing will be very difficult to implement.

The underlying reason for utilizing optical fiber within a cabling structure is to provide communications paths for an extended period of time without replacement. This contrasts strongly with the *ad hoc* cabling approach adopted in many areas to service frequently changing communication needs where copper cabling is easier to replace, repair and reconfigure. Examples of this approach are to be seen in telecommunications as the subscriber loop and in data communications as the office floor area.

That being said, optical fiber is now finding its way into the home and onto the office floor as the technology cost drops and a better understanding of future traffic needs are identified.

In any case the use of optical fiber within a cabling structure places a responsibility upon the user for defining the present and future services required by the users. It also places a responsibility upon the designers and installers to produce a cabling structure which meets these requirements and can be proven to do so. Finally, a future-proof communications medium must be seen to produce cost benefits to its users. Correct analysis of these requirements is the aim of all concerned and the remainder of this book concerns itself with the subject of highway design, installation practice and highway operation.

Optical fiber highway design

The design of the optical fiber highway should take account of the transmission equipment to be used both initially and in the future. The latest trend towards networking standards gives some pointers with regard to transmission technology and the wider impact of commercial pressures cannot be ignored.

The topology of the highway should also reflect the future expansion of services to existing locations and the introduction of services to new locations.

The repairability philosophy will further influence the final layout of the cabling. All these issues are covered in Chapter 9.

Component choice

Once the basic design is established it is necessary to assess the components to be used within the design.

The fixed cables, jumper cables, patch cables, connectors and the jointing techniques must be chosen to meet the environmental requirements of the installation.

In addition, the specifications of the components must be defined in order to ensure that the design adopted will function for the proposed life of the optical fiber highway. These issues are discussed in Chapter 10.

Specification agreement

It is remarkable, not to say amazing, that the total specification for the installation of an optical fiber highway can still be found in an invitation to tender as

> Please supply a fiber optic network.

The resulting submissions can vary from the sublime to the ridiculous, with pricing to match. The construction industry, with much greater contractual experience, uses highly detailed specifications to ensure that the description of the task and the implications for the installer and user alike are well understood. An optical fiber highway is, in most cases, built to last and for this reason a specification is equally desirable. The production of a specification is covered in Chapter 11.

Component testing

The trouble-free contractual management of an optical fiber highway installation is dependent upon correct specification of the components to be used, the assessment of the components against their specification and finally the quality of the installation workmanship.

Failure to inspect incoming goods can lead to delays. In capital projects where the time scales are critical, any delay can lead to lost revenue, damage to reputation and severe disruption. Acceptance test methods are defined in Chapter 12.

Installation practices

While the practices adopted for the installation of optical fiber cable do not differ in most respects from those used in copper cabling there are a few aspects which merit detailed explanation. This is covered in Chapter 13.

Final highway testing

The subject of testing the installed highway is complex and comparisons between measured results have been misleading.

Chapter 14 highlights the issues of measurement and relates them to the specification of transmission equipment.

Optical fiber highway documentation

The optical fiber highway is not a static solution to a problem. Rather it is a cabling structure which may be reconfigured and extended at will. Modifications may not necessarily be made by the original installer and over the lifetime of the highway the users will come and go.

Full and detailed documentation is therefore a necessity. Chapter 15 introduces the concept of the organic highway and defines the documentation necessary to service the growth and spread of the organism.

Repair and maintenance (and user training)

The design of an optical fiber highway must take account of the repair philosophy to be adopted with a view to minimizing the operational downtime of the services being offered on the highway. Nevertheless the cabling structure will eventually become defective and it is important to know how to allocate responsibility, locate the fault and effect a high-performance repair.

Much can be achieved by effective training of the user and Chapter 16 details the type of training that should be given.

9 Optical fiber highway design

Introduction

An optical fiber highway can take an infinite variety of forms, ranging from a single point-to-point link, directly connected to equipment, up to a large multi-node, multi-element, multiple fiber geometry structure with both terminated and unterminated (dark) fibers. The final design is, in all cases, aimed to provide the desired level of expansion, evolution, reliability and reparability.

The design can be divided into the following parts:

- *Highway topology* The fixed cabling layout
- *Nodal design* Active and passive node configurations
  Patching facilities
  Repair philosophy
- *Service needs* Current service requirements
  Future services and standards
  Fiber count and fiber geometry

To some extent the design is iterative and may be addressed in a number of ways. Real expertise in optical fiber highway design can really only come from experience but the basic issues are not complex. This chapter seeks to spread the understanding and underline the desirability of good design.

Highway topology

It should be pointed out at the outset that the desirability of good design is ever present. It is not confined to large cabling structures such as those seen in telecommunications and campus style networks and is equally important in short-range network solutions as used in aircraft, ships and other fighting vehicles.

Highway topology is an all-encompassing term which defines the permanent nature of the highway and includes the provision of optical fiber to all desired nodes, be they passive or active (or a mixture of the two).

Figure 9.1(a) shows a typical campus style application. A total of 11 buildings are grouped in a seemingly random pattern around a site. Each of the buildings may contain one or more nodes in a communication system.

Figure 9.1(b) shows a typical backbone style application. A total of six floors in a building create a vertical structure in which each of the floors could house one or more nodes in a communication system.

Figure 9.1(c) shows a typical high-connectivity application often seen in short-range networks such as military vehicles, where compartmentalization of the vehicle (by the use of bulkheads) leads to the concept of nodes (at the transmission equipment) and subnodes (at the intervening interfaces).

A realistic highway might feature combinations of the above with a large fixed cable content comprising both external and internal sections reflecting both campus and backbone configurations.

Each of the three examples of campus, backbone and high-connectivity environments has to be treated slightly differently from the design viewpoint.

Campus style topologies

The campus is rapidly becoming the most common environment for optical fiber in the data communication market sector. The primary reasons for the use of the optical medium in these applications are summarized below:

- The high bandwidth medium enables the inter-building cabling infrastructure to be installed once only. Any subsequent service changes and upgrades may be achieved by the replacement of transmission equipment and the modification of patching facilities. The spread of the optical solution into the buildings occurs as and when required.
- The insulating properties of the optical fiber provide an electrical isolation between the nodes. Differentials in earth potentials between buildings on the campus are therefore not a problem. Similarly the possibility of damage resulting from lightning discharge through the cable is drastically reduced. That being said the metal content of the fixed cables (strength members and moisture barriers) and the methods of fitting within the termination enclosures merit careful attention to maximize this benefit.

Obviously every requirement is different and the additional benefits of security of transmission, extended distance of transmission (due to the low

160 Fiber Optic Cabling

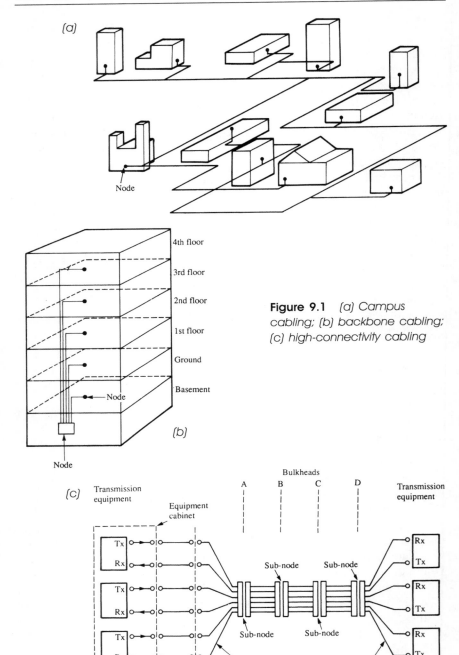

Figure 9.1 (a) Campus cabling; (b) backbone cabling; (c) high-connectivity cabling

signal attenuation exhibited by optical fiber) and electromagnetic signal immunity can further influence the decision in favour of optical fiber.

The ideal, but frequently unacceptably expensive, requirement is for direct communication between each and every node. This is obviated by the use of networking protocols such as Ethernet and ATM. The need to directly link every node can therefore be reduced by analysing the campus as a series of clusters which must themselves be connected. The alternative is to view the campus as a ring structure.

Although this is not intended to be a text on communication standards it is worthwhile explaining the basic optical configurations used to interconnect Ethernet and token ring networks.

Except for highly specialized custom-built solutions, there are no fiber optic bus networks. In general a copper-based Ethernet bus must be terminated with a fiber optic transceiver. This transceiver is then connected via the highway to another fiber optic transceiver (which is connected to another copper-based Ethernet bus) or to a fiber optic repeater which connects to a number of other fiber optic transceivers.

It should be pointed out that, in general, the transceivers communicate on two separate optical fibers (one transmitting, the other receiving). Bi-directional transmission on a single optical fiber is possible but is comparatively expensive. However, as integrated optics technology improves, the chip-based combination of optical source and detector may open up possibilities in this area.

An optical fiber token-passing ring, such as IEEE 802.5 token ring or FDDI, is configured in the same manner as its copper equivalent. Each node contains two fiber optic transceivers, one operating clockwise around the ring and the other operating anticlockwise.

Any failure within the ring (either as an equipment or cable defect) is resolved by a reconfiguration. In certain cases a contra-rotating or dual redundant ring is used.

For any campus site the infrastructure may be viewed as consisting of a number of key nodes (each of which is surrounded by a cluster of lesser nodes) with the key nodes connected to each other in some type of star arrangement. This is the basis of an Ethernet configuration.

Optical fiber versions of Ethernet are 10BASE-F, 100BASE-FX, 1000BASE-SX and 1000BASE-LX. A recent 100 Mb/s, 850 nm version is called 100BASE-SX. 10BASE-F and 100BASE-FX (full duplex) can transmit up to 2000 metres over multimode fiber.

Gigabit Ethernet over optical fiber has yet another set of rules. Unlike previous optical LAN transmission systems, gigabit Ethernet is bandwidth limited. Most other systems are attenuation limited. Three different types of optical fiber are allowed: 50/125, 62.5/125 and single mode. Two different bandwidth grades are supported within each of the two multimode fiber styles. The quality of the fiber is determined by its available

Table 9.1 *Optical gigabit Ethernet requirements as per IEEE 802.3z*

Fiber type	Fiber bandwidth MHz.km at 850 nm	at 1300 nm	Transmission distance at 850 nm	Transmission distance at 1300 nm
62.5/125	160	500	220 m	550 m
62.5/125	200	500	275 m	550 m
50/125	400	400	500 m	550 m
50/125	500	500	550 m	550 m
Single mode				5000 m

bandwidth, and Table 9.1 demonstrates the link lengths possible with the different fiber types.

Ethernet continues to evolve and now the next generation is ten gigabit Ethernet, or 10GBASE-xyz, where:

- x = S (short wave, 850 nm), or L (long wave, 1300 nm) or E (extra long wave, 1550 nm);
- y = W (WAN using SONET ST-192 encoding) or R (LAN using serial encoding) or X (LAN using CWDM encoding);
- z = the number of CWDM channels.

CWDM means coarse wavelength division multiplexing, where the space between wavelengths used is much greater than would be encountered in telecommunications dense wavelength division multiplexing or DWDM. In the above scheme, 10GBASE-LX4 would mean ten gigabit Ethernet working at 1300 nm with four wavelength division multiplexed channels.

Ten gigabit Ethernet will be described in the standard IEEE 802.3ae and will be completed by March 2002. The philosophy and justification of another factor of ten increase in speed remains the same. If numerous users are generating data at 1 Gb/s or even many users at 100 Mb/s then even a 1 Gb/s backbone will soon become overloaded.

Ten gigabit Ethernet over single mode fiber is relatively trivial, as this is a current telecommunications speed. Several other methods are up for discussion with the aim of getting at least 300 metre transmission distance over multimode and tens of kilometres over single mode. Unfortunately existing or legacy multimode fiber does not have the bandwidth to cope with a straightforward 10 Gb/s data stream sent down it. Figures of between 28 and 86 metres transmission distance have been suggested if legacy fiber should be used this way. This would be for multimode working at 850 nm with a VCSEL laser. One hundred metres should be obtainable with a Fabry Perot laser working at 1300 nm. The theoretical legacy fiber options are:

- Serial coding with 850 nm laser on 50/125 legacy fiber   86 metres
- Serial coding with 1300 nm laser on legacy fiber   100 metres
- Parallel optics, i.e. 2.5 Gb/s sent down four separate fibers 300 metres
- Wavelength division multiplexing, i.e. 2.5 Gb/s sent down the same fiber but using four different wavelengths or 'colours' of light   300 metres

Another option is to introduce a brand new multimode fiber, laser launch optimized and with a much higher bandwidth. This would give a new 50/125 fiber a 300 metre range with the low cost VCSEL.

The final option is of course to use single mode fiber. There will be 1300 nm options for a few kilometres range and 1550 nm options for at least 40 kilometres range.

Local area networks were developed to communicate between general-purpose servers and workstations, i.e. the client–server model. Although LANs are very good at connecting large amounts of users they are not optimized for transferring large quantities of data, especially between mainframes. Links between mainframes and their high-speed peripherals are called 'channels' in this context. We thus have a family of channel protocols that do a different job from local area networks.

Channel protocols, such as SCSI and Bus & Tag, started off as short-distance links but have evolved into longer-distance, very high-speed channels such as ESCON and Fiber Channel. The short-distance channels such as SCSI and the old IBM Bus & Tag require dedicated copper cables but the optical links will be expected to run over the same optical backbone cabling as any other optical LAN. The cable network designer must therefore take into account the cabling requirements of LAN backbones such as FDDI, 100BASE-FX. 1000BASE-LX, 1000BASE-SX, 10GbE and ATM etc., along with the channel requirements of systems such as ESCON, HIPPI and Fiber Channel. Table 9.2 summarizes the current offerings.

Careful consideration should be given to the choice of nodes. In the campus situation every building could be regarded as a node but there are limitations. For instance, it is highly unlikely that the site canteen may ever become a key communications site and therefore it probably is not necessary to visit that location with the cabling infrastructure. Nevertheless there will be locations that whilst not needing connection to the infrastructure immediately may eventually require some level of service. These buildings should be visited by the optical fiber highway and either left with a service loop of cable to enable future commissioning or alternatively commissioned during the initial installation and jointed through to minimize the signal attenuation at that node.

If there are entire clusters of nodes which are initially to be non-operational, then the cabling installation may be phased to reduce the

Table 9.2 *Optical LANs and channels*[1]

Channel	Speed
ESCON	136 Mb/s
Serial HIPPI	800 Mb/s
HIPPI-6400 (GSN)	6400 Mb/s
Fiber Channel	133 to 1062 Mb/s

Optical LANs	Speed
Ethernet 10BASE-F	10 Mb/s
Ethernet 100BASE-FX	100 Mb/s
Ethernet 1000BASE-SX/LX	1000 Mb/s
Ethernet 10GbE	10 000 Mb/s
FDDI	100 Mb/s
ATM	155 to 2400 Mb/s
Token ring IEEE 802.5v	1000 Mb/s

immediate capital cost. The cabling of the cluster (or ring segment) can take place later but provision must be made at the time of the initial installation to enable connection of the future cable with the least cost and, probably more importantly, minimal network disruption.

Figure 9.2 shows the ISO 11801 hierarchical, three-layer campus topology. Optical fiber can appear as the horizontal cabling, up to 100 metres, the building backbone, up to 500 metres and/or the campus backbone up to 1500 metres. The building and campus backbones can be linked to form one 2000 metre span. The 2nd edition of ISO 11801 caters for three new optical channels, OF300, OF500 and OF2000 as well as centralized optical architecture which is covered in more detail later.

This section has discussed the issues of topology or layout of the cabling infrastructure for a campus style site. The next section deals with the installation topologies internal to buildings.

Building backbone topologies

The utilization of optical fiber within buildings is a growth market. The reasons for its use are broadly in line with those put forward for the campus style infrastructure; however, the additional benefits of low cabling mass and volume can also influence the decision to opt for optical fiber. The backbone is normally assumed to be vertical (although it doesn't have to be) and is installed in the vertical risers of buildings.

Optical fiber highway design 165

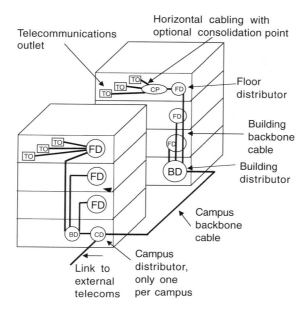

Figure 9.2 *ISO 11801 campus cabling model*

Each floor is regarded as a node and the cable infrastructure provides data communication up and down the backbone, allowing each floor node to transmit and receive to and from every other floor node.

The use of optical fiber past the nodes onto the floor (to the desk) is a subject in its own right and the economics of such a migration of optical technology is discussed later in this book.

The standards of communication are the same as those in the campus environment, although possibly at a lower speed. The difference between campus and backbone topologies is that within a building there tends to be a central location acting as the communications centre. This is often the main computer room and although it is most frequently found in the basement it can be located anywhere within the building. This focal point acts as a concentration point for all the patching facilities needed by the entire building.

As a result the topology adopted tends to be a star with each floor being individually serviced from the focal point. This infrastructure is independent of the star or ring requirements of the communication standard and is much more related to the desire to route directly between floor nodes on a point-to-point basis using a large patching field.

This topology is shown in Figure 9.3. This type of star-fed backbone in no way conflicts with the ring infrastructure which may be used for

166 *Fiber Optic Cabling*

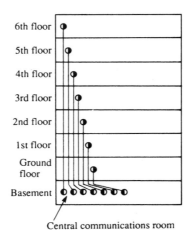

Figure 9.3 *Star-configured backbone*

the campus application since in both cases it is the infrastructure that is being discussed and not the networking standards. This underlines the difference between an optical fiber highway and the network services for which it is configured.

Redundancy and reparability

The highways designed for campus and backbone applications are analogous to the trunk cabling within the telecommunications industry. They are responsible for the transmission of large amounts of information between buildings or between floors within buildings. Once installed the reliance upon, and utilization of, the highway will increase to the point where downtime will be considered to be costly. The only method of establishing the cost of network downtime caused by highway failure lies in the hands of each customer. The cost of failure of a vital link in a major clearing bank may be significantly higher than a similar failure in a non-essential communication link in a university but the only source capable of assessing that cost is the user. As the estimated cost of failure rises then so does the necessity of designing the highway in such a way as to limit or even eliminate downtime.

Practical solutions rely on a combination of highway resilience (through the use of dual redundant cabling or spare fiber elements), equipment resilience (through the use of dual redundant equipment or physical reconfiguration) or protocol-based resilience (through automatic software-based reconfiguration).

If fully dual redundant highways are to be used then the cost of the highway will virtually double since a mirror image has to be installed.

This is particularly true of campus style highways where, in order to produce full dual redundancy, the two cables must be laid on separate routes and brought into the nodes at different points.

Within backbone highways the existence of a large patching facility can, under the right circumstances, achieve adequate protection by using separate risers and having a secondary communications centre which could provide support should the central communications room be damaged in some way.

High-connectivity highways

Campus and backbone applications account for the vast majority of the data communications usage of optical fiber. The majority of military applications also involve these cabling infrastructures (e.g. within government and defence-related establishments). However, the most costly installation per metre of cable used must be those in the various types of fighting vehicles and aircraft.

These are very short-range highways; in many cases less than 100 metres in total length. The installation of these systems, for in many cases the highway serves only one proprietary purpose and is therefore part of the transmission system rather than a universal communications medium, is undertaken in radically different conditions to those found in the campus and backbone environments. Also the methods used to test both the components and the final cabling are radically different from those used on the longer, system-independent cabling structures found in campus and backbone applications.

In both the campus and backbone environments the task is to install a design, whereas the short-range high-connectivity highways require the installation itself to be designed. This has direct consequences for the topology of the optical fiber highway.

High levels of connectivity result from:

- The existence of physical barriers such as bulkheads which compartmentalize the installation.
- The need to repair by replacement rather than rejointing or insert-and-joint techniques.

The transmission equipment may be separated by multiple demountable connector pairs and these can be thought of as subnodes. Therefore each pair of nodes is connected through a number of subnodes and between each pair of subnodes there runs a directly terminated cable assembly.

The compartmentalization of the highway limits the use of termination enclosures due to their size and the need for easy replacement of the cable assemblies. Accordingly the cable assemblies act as jumper or patch cables between the subnodes. The assemblies are not fixed cables as are the internodal links in the campus and backbone environments but are fixed cable looms as described in Chapter 7.

These patch cables may take the form of multiple SROFC (single ruggedized optical fiber cables) units each of which is directly terminated and subsequently loomed together. Alternatively larger multi-element cables may be terminated with one of the multi-ferrule connectors currently available.

The use of multiple demountable cable assemblies emphasizes the need for careful design since the connector pairs tend to be a source of damage and other reliability problems. Highway failure will eventually be repaired by the replacement of the damaged assemblies but this will take time and a topology must therefore be adopted which makes possible a rapid resurrection of the highway without attempting to repair the damage. This topology may include dual redundancy with separate looms taking separate routes or spare elements within the cable assemblies, which can be used following reconfiguration at the transmission equipment.

Nodal design

The design of the optical fiber highway at the nodes is one of the most important features within the overall design. The flexibility of the infrastructure is determined by the node design but flexibility is normally achieved at the expense of signal loss. As a result the ultimately flexible highway design can be inoperable. There are three possible nodal designs that are discussed in this section.

Active and passive nodes

The design of the ultimate highway would include direct connection to every optical fiber at every node. This would ensure total flexibility in the configuration of the highway for the various networked services to be offered to the users. However, the losses at demountable connectors can in many cases be equivalent to many hundreds of metres of installed cable and it makes a great deal of sense to create a design which combines the desired (rather than the ideal) degree of flexibility with acceptable levels of signal attenuation. To this end it is logical to provide optical input and output only at nodes in which communications will actually be needed.

As time passes the nodes needing physical access to the highway will change and the number of communicating nodes will tend to increase. It has already been said that from an economic viewpoint it is sensible to provide cabling to all sites (be they buildings on a campus or floors in a building) independent of their initial needs.

This then develops the concept of the active node (one with physical access to the highway) and the passive node (one with no physical access but with provision to gain access at some future date).

Passive nodes

The passive nodes fall into two categories. The first is one in which the fixed cabling merely visits the location. Spare cable in the form of a service loop is left in a convenient position for future connection.

The second is normally to be found when it is deemed likely that the location will require services in the medium term yet the services required are not defined. In these cases the fixed cabling enters a termination enclosure and is secured, the optical fiber prepared and marked and subsequently jointed (by either fusion or mechanical methods) such that the communication between the nodes on either side can take place.

A typical termination enclosure may comprise a 19-inch subrack unit fitted with a set of glands, strain-relief mechanisms and cable-management facilities. Alternatively wall-mounted boxes are used where space or other requirements render a 19-inch cabinet undesirable. There are many designs of termination enclosure available but few standard solutions. The recommended requirements of the generic termination enclosure are detailed later in this book.

From the point of view of the optical fiber highway, nodes are either passive or active depending upon whether physical access to the highway can be gained. The active node designs that follow cover the entire spectrum of those seen in real installations.

Active direct terminated nodes

A direct terminated node is one in which the incoming cables from other nodes are terminated without the use of a termination enclosure. An example of this is shown in Figure 9.4, where because of space restrictions it has been decided to use directly terminated cables at one end of the fixed cables. The cables could be terminated on site or alternatively the fixed cable could be preterminated at one end and installed from that end (so as not to damage the terminations).

The apparent advantage of this method is cost since it would seem that the elimination of the termination enclosure would reduce installation price. However, the method can only be used in a professional manner on cables containing breakout SROFC units, which can become costly as the fiber count increases. Real cost savings can therefore only be found when the number of optical fibers is low (normally only two).

The disadvantage of this technique is that damage to one of the terminations within a duplex link renders the link useless unless a spare cable element has been terminated. This presents difficulties with storage of the spare element and it is rare that such a configuration would be adopted

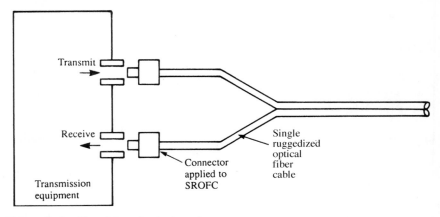

Figure 9.4 Direct terminated node

in practice. As a result a directly terminated node is used only when failure of the highway is regarded as non-critical.

Active pigtailed nodes

When the fixed cables are critical to highway operation, each internodal link needs to be fitted with a termination enclosure at either end. The termination enclosure has a dual role: it allows effective strain relief for the fixed cable (and the optical fibers within it) and it provides protection for the optical fibers within the cable.

Termination enclosures can be configured in two ways: pigtailed or patch. The pigtail option is designed such that the fixed cable is glanded into the termination enclosure, the optical fibers are prepared, marked and then SROFC elements are either fusion or mechanically spliced to them. The SROFC elements are preterminated at one end only with a connector and such a cable is called a pigtail. (To be accurate the terminology is rather that the connector is said to be pigtailed.)

The pigtail is glanded, i.e. secured with a cable gland into the termination enclosure (with strain relief provided by means of the yarn-based strength member within the SROFC construction), and leaves the termination enclosure for onward connection to the transmission equipment. The connector is described as the equipment connector. An example of this design is shown in Figure 9.5.

An additional benefit resulting from the use of termination enclosures is that spare optical fibers within the fixed cable can be jointed to pigtails and stored safely within the enclosure. Should damage occur to an operating pigtail during normal use, the spare element can be removed from the termination enclosure, connected to the equipment and network

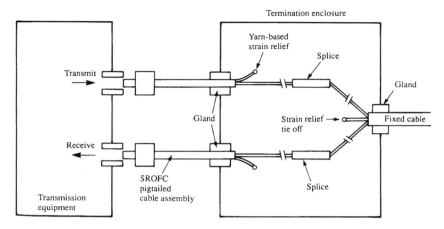

Figure 9.5 *Pigtailed termination enclosure node*

operation restored. The damaged pigtail can be placed in the enclosure pending a full repair. This can be effected without disruption to the highway.

The disadvantage of the pigtailed node design is that new equipment with a different equipment connector will require the rejointing of a new pigtail within the termination enclosure. A much more flexible configuration is the patched node as described below.

Active patched nodes

In a patched node the termination enclosure is fitted with a panel containing demountable connector adaptors. The fixed cable entering the termination enclosure is glanded, the optical fibers prepared, marked and then jointed (by either fusion or mechanical splice) to SCOF pigtails. The connector on the pigtails is inserted into the rear of the panel adaptors, thereby creating a patch field. This is shown in Figure 9.6. Alternatively, the optical fiber with the fixed cable may be directly terminated on-site and fitted into the rear of the panel adaptors. This design of node has all the advantages of the pigtailed node but benefits from network flexibility. The patch node connector is termed the system connector and normally reflects the latest technology and standards within the demountable connector marketplace. It is isolated from the equipment and is not influenced by equipment choice. A change of equipment merely requires the purchase of an appropriate jumper cable with the system connector at one end and the equipment connector at the other. It is undoubtedly true that the majority of users would prefer patch nodes on all occasions but their indiscriminate use can create optical losses which exceed the

172   Fiber Optic Cabling

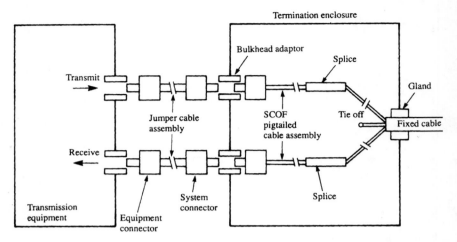

Figure 9.6   Patched termination enclosure node

capability of the proposed transmission equipment. This is discussed later in this chapter.

The final choice of node design depends not only upon the desired levels of reparability and flexibility within the highway but also upon the network services to be provided (on a node by node basis). As a result it is not uncommon to find an active node featuring both pigtailed and patch facilities. Similarly an active node (as defined at the highway level) may feature passive node elements at the network level. This network services approach is covered in the next section.

Service needs

Current needs

The optical fiber highway has been defined as that infrastructure of fixed cabling linking the nodes within a proposed communication network. The highway is subsequently configured according to the specific services which will be offered to the users within those nodes.

The choices with regard to the treatment at the nodes of the optical fiber within the fixed cabling are made after due consideration has been made as to the services to be accessible at that node.

When true networking is desired (that is when every operational node can communicate directly with every other by means of a standard communications protocol such as Ethernet) it is obviously necessary to provide highway access at participating nodes. However, when nodes exist for reasons of geography (at concentration points on the cable routes) and no communications equipment is intended to be situated within them it is sensible to treat them as completely passive nodes and there is little need for access to the highway.

Similarly the provision of point-to-point services suggests that intermediate nodes should be treated as passive (with no optical access) for the optical fibers involved.

The recommendation therefore is to adopt a passive node design for all fibers included in the initial service requirement except for those optical fibers at those nodes which need access to the highway for reasons of communications or flexibility. The jointing-through of optical fibers at nodes has a number of operational advantages:

- *Low attenuation*
  Whether fusion or mechanical splice techniques are used the attenuation introduced is generally lower than for the patch alternative. This is particularly important for point-to-point services which must traverse many intermediate nodes.
- *Reliability*
  Demountable connectors are always a source of potential reliability problems largely due to contamination introduced by careless handling. It is important to minimize the number of demountable connections by the use of permanent joints where possible. This must be balanced by the need to maintain flexibility by the use of patching facilities with controlled access.

The above recommendation suggests that all fibers within the highway should be jointed at nodes at which there is no definite access requirement. Upon re-reading it will be seen that this only applies to those optical fibers included in the original operational specification. There may be a significant number of additional elements within the fixed cabling of the highway which are not required in the initial configurations of the highway. These can be left unjointed and unterminated, commonly termed 'dark fibers'.

Future needs (campus and backbone applications)

It is always difficult to predict the future requirements for communications between users. Unfortunately in order to maximize the return on capital invested in the cabling infrastructure it is necessary to attempt to analyse future needs by looking at the present technological trends.

A given cabling infrastructure can become saturated (i.e. no further expansion of services can be achieved) when either:

- the number of individual services to individual users exceeds the physical provision of the infrastructure, or
- the communication rate necessary to provide the services exceeds the bandwidth of the communication medium.

The first of these limitations is the most difficult to analyse. A typical campus environment may include buildings in which there are existing copper-cabling infrastructures which support Ethernet and other perhaps less well-known systems. If these are to communicate with their fellows around the campus some provision must be made in terms of allocating optical fibers to these services. Alternatively decisions may be made which standardize upon a particular network carrier over the optical fiber highway. For instance, all the locations may communicate over optical Ethernet provided that the correct bridging or interfacing equipment is available for the existing systems within the buildings.

However, there are certain services that do not lend themselves to becoming part of a network due to their bandwidth requirements or the nature of their interconnection. Examples of these are high-speed services such as high-resolution video, RGB, CADCAM and secure point-to-point services as may be needed to be separated from any conventional and accessible network structure. Users are likely to ask for these applications-specific communications but predictions of likely uptake are notoriously difficult to make. As a result there is no realistic factor of safety which can be applied to optical fiber highways and each one has to be taken and assessed on its own merits.

Nevertheless some guidance can be given as to a minimum provision. At the time of installation of the optical fiber highway the networking requirement may be considered as justifying the inclusion of a number of optical fiber pairs. The total number of fibers thus calculated represents the base requirement to service the networking needs. Merely to install this number would be irresponsible and could almost be guaranteed to necessitate a further cable installation in the short to medium term (thereby totally defeating the purpose of the optical fiber highway). The known quantity of applications-specific services should be added to this base requirement and the result doubled. An installed cable containing this quantity of optical fibers (as a minimum) should be able to provide an adequate level of future-proofing.

An example of this calculation could be a site comprising eight individual buildings requiring to communicate on both ATM and Ethernet. In addition a point-to-point Fiber Channel service is required between two sites for main processor communication and a video surveillance link service has been requested between two other nodes.

In this case two pairs of fibers are needed for networking purposes and a further pair is needed for the main processor link. Video surveillance is assumed to require only a single fiber; therefore all seven fibers are needed immediately. Doubling this to 14 would therefore seem a sensible precaution against recabling. A standard cable size is 16 fibers, and this is the most appropriate cable selection for this project.

The doubling allows for service expansion but is also in keeping with the philosophy of maintaining at least one spare fiber for repair and basic reconfiguration purposes. However, it is important to underline that there is no true fiber count calculator–predictor, and as many optical fibers as possible should be included within the fixed cabling content of the highway.

It should be remembered that optical fiber is inherently of low unit cost and that a large unterminated fiber content will not necessarily significantly impact the overall installation cost.

Research by the author has revealed some facts about the number of optical fibers currently being put into installations across Europe. The average number of fibers (on those sites with optical fiber) was 201. This seems a surprisingly high number, but the average is skewed by a small number of fiber-to-the-desk projects with a vast number of fibers to the desk (17 000 being the maximum in this survey). In this case it is more instructive to use other statistical tools such as the median and mode. The median is 24 and the mode is eight. The 'real' average therefore is eight, and this is borne out by manufacturers' shipping data. The full survey details are in Table 9.3.

Table 9.3 *Survey of optical cable in LAN installations*

Number of sites	459
Location	Europe
Dates of installation	1996–2000
Total number of fibers installed	92 284
Maximum	32 376
Minimum	1
Average	201
Median	24
Mode	8

Having discussed the number of fibers in an effort to ensure that the service requirement will not outstrip capacity it is now relevant to discuss the data-carrying capacity of the individual optical fibers.

Optical fiber technology is, to some extent, a paradox. The optical fiber with the greatest bandwidth capable of carrying almost limitless amounts of data is also the cheapest product in the range. All the other fiber geometries have more limited bandwidths and are more expensive to produce. It would appear sensible therefore to install single mode fibers in every cabling project in the firm knowledge that the highway will never become jammed with competing information. Unfortunately the situation is not clear-cut and deserves extended consideration.

An optical fiber highway consisting of a number of spans can fail for two primary reasons:

- insufficient optical power being received by an optical detector;
- corrupted information being received by an optical detector.

The first suggests that careful design be undertaken with regard to the amount of power coupled into the optical fiber by the transmission equipment and a strict assessment of attenuation be made to ensure operation under all circumstances. The second suggests that the optical fiber used shall have the maximum bandwidth consistent with the constraints of the first.

To understand the choices open to the designer of the optical highway it is necessary to set out some of the basic definitions within an active optical span.

Optical budget

Figure 9.7 shows the basic construction of a basic optical fiber communications subsystem. At one end of the optical fiber is an optical transmitter and at the other there is an optical receiver. In reality each of these opto-electronic units is housed in a black box and access is gained to them via a chassis-mounted receptacle (Figure 9.8(a)) or via a connectorized pigtail (Figure 9.8(b)).

The optical transmitter has a product specification and part of that specification is the optical output power. As the amount of optical power coupled into the various fiber geometries will differ (due to core diameter and NA differences) then the optical output power must be specified as a function of the fiber geometries into which the transmitter may be connected.

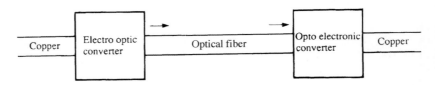

Figure 9.7 *Basic optical fiber communication subsystem*

Optical fiber highway design

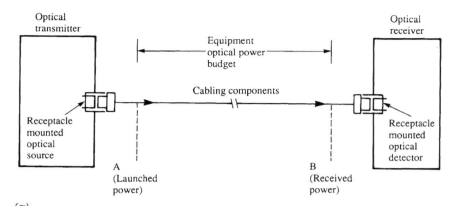

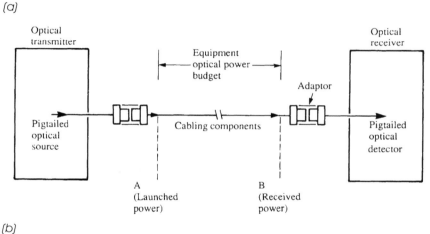

Figure 9.8 *(a) Chassis-mounted receptacle-based equipment; (b) connectorized pigtail-based equipment*

As an example a first window transmitter might be quoted as launching the following optical powers:

Launch fiber   *Power coupled*
100/140 μm 0.29 NA   0.1 mW = 100 μW = −10 dBm
62.5/125 μm 0.275 NA   40 μW = −14 dBm
50/125 μm   0.20 NA   16 μW = −18 dBm

It should be pointed out that the variation in coupled power will not necessarily be in accordance with the core diameter and numerical aperture mismatch equation shown earlier in this book. Taking the 100/140 μm 0.29 NA fiber as a standard the equations predict a reduction of 4.6 dB (62.5/125 μm 0.26 NA) and 9.3 dB (50/125 μm 0.20 NA) being launched into the respective fiber types. The flaw in this argument is that it assumes that the light distribution from the optical source into the optical fiber is uniform with an NA of greater than or equal to 0.29. Many devices do not operate in this way and tend to concentrate optical power towards the axis of the optical fiber. For this reason the powers launched into the smaller-core, lower NA geometries may be higher than calculation might show. It is important therefore to obtain from the transmitter manufacturer a written specification for the powers launched into the various fiber geometries rather than using a calculated figure.

The powers coupled will have a maximum and a minimum value for a given fiber geometry. The maximum power coupled will be applicable to a brand new device and will be measured at a given temperature. For a light emitting diode (LED) the maximum output will be found at the lowest operating temperature whilst for a laser source the maximum output will be achieved at the highest operating temperature. Similarly the minimum output power for an LED source will be found at the highest operating temperature whilst for a laser this figure will be measured at the lowest operating temperature. In general LED sources decrease in efficiency during their operating life and this ageing effect must be included in calculating the minimum power coupled into any fiber geometry (normally assumed to be -3 dB). Lasers, however, tend to be stabilized via feedback circuitry and such an ageing loss is not always necessary.

To summarize, a transmitter will be specified with a maximum and a minimum power coupled into a given fiber. The user may wish to investigate the published values to ensure that all of the above factors have been taken into account within the data provided. Finally it should be ensured that the manufacturer has measured this optical power at the point A in Figures 9.8(a) and (b) via an acceptable connector joint.

The power measurements relevant to the detector are more straightforward. The detector consists of a silicon (first window) or germanium (second window) photodiode which will react when optical power of the correct wavelengths is incident on its surface. It is normal for the photodiode to be larger than the optical core and any direct connection is assumed to exhibit negligible attenuation. Any pigtailed connection is taken to be part of the receiver package and therefore measurements of optical power are taken at point B in Figures 9.8(a) and (b).

Any detector will have a maximum input power (at which the signal saturates the detector/receiver circuitry causing errors in reception and/or possible damage to the equipment) and a minimum input power below which reception is not guaranteed and errors in transmission result.

For instance, a first-window detector might be specified as follows:

maximum received power: 40 µW = -14 dBm
minimum received power: 1.6 µW = -28 dBm

These levels of input power tend to be independent of optical fiber geometry and do not vary with temperature or age.

To summarize, a detector will be specified with a maximum and a minimum input power or sensitivity. When analysed in conjunction with the transmitter specification it allows an assessment of the optical fiber geometries to be used in terms of attenuation allowed. In this way an optical power budget can be produced for each optical fiber geometry.

There are many ways of demonstrating an optical power budget calculation including graphical means; however, it is quite adequately analysed by addressing each possible optical fiber geometry in turn. Taking the example of the above transmission equipment it is possible to make some revealing statements about the allowable span designs.

Table 9.4 *Transmitter coupled power*

Fiber		Maximum	Minimum
100/140 μm:	Optical power coupled	−9 dBm	−11 dBm
	Temperature allowance		
	−10°C	+1 dB	
	+50°C		−2 dB
	Ageing allowance		−3 dB
		−8 dBm	−16 dBm
	Optical power received	−14 dBm	−28 dBm
	Minimum link loss	6 dB	
	Maximum link loss		12 dB
62.5/125 μm:	Optical power coupled	−13 dBm	−15 dBm
	Temperature allowance		
	−10°C	+1 dB	
	+50°C		−2 dB
	Ageing allowance		−3 dB
		−12 dBm	−20 dBm
	Optical power received	−14 dBm	−28 dBm
	Minimum link loss	2 dB	
	Maximum link loss		8 dB
50/125 μm:	Optical power coupled	−17 dBm	−19 dBm
	Temperature allowance		
	−10°C	+1 dB	
	+50°C		−2 dB
	Ageing allowance		−3 dB
		−16 dBm	−24 dBm
	Optical power received	−14 dBm	−28 dBm
	Minimum link loss	nil	
	Maximum link loss		4 dB

The specification of the transmitter coupled power is expanded in Table 9.4.

It can be seen that the maximum and minimum attenuation allowable in the spans varies with the optical fiber to be used. This therefore defines the level of complexity in terms of connectors, joints and fiber length which can be accommodated within the span.

The results of the above analysis may be summarized in Table 9.5.

Table 9.5 *Optical power budget calculations*

Fiber geometry	Optical power budget	
	Min (dB)	Max (dB)
100/140 μm	6	12
62.5/125 μm	2	8
50/125 μm	nil	4

This suggests that the cabling between points A and B in Figures 9.8(a) and (b) must lie within these boundaries, otherwise the equipment may not perform to its full specification. Some allowance must be made for future repairs (perhaps 1 dB) and therefore the link can be configured with specified losses up to 11 dB, 7 dB and 3 dB in the respective fiber geometries. Unfortunately it is frequently seen that the manufacturer of the transmission equipment will specify performance over one fiber

Table 9.6 *ISO 11801 optical component loss allowance*

	850 nm multimode	1300 nm multimode	1310 nm single mode		1550 nm single mode	
			Inside plant	Outside plant	Inside plant	Outside plant
Optical cable						
ISO	3.5	1.5	1.0	1.0	1.0	1.0
TIA	3.5	1.5	1.0	0.5	1.0	0.5
Optical connectors	0.75 dB	0.75 dB	0.75 dB	0.75 dB	0.75 dB	0.75 dB
Optical splice	0.3 dB	0.3 dB	0.3 dB	0.3 dB	0.3 dB	0.3 dB

Note: all cable measurements are dB/km

All figures are the same for ISO 11801 2nd edition and TIA/EIA 568B except where identified with '*TIA*' or '*ISO*'

Optical fiber highway design 181

geometry only. This makes the data sheet more simple and certainly assists a non-specialist salesman in streamlining the approach with the potential customer but it does not do justice to either the product or the user.

For the purposes of link assessment the maximum component losses are specified in ISO 11801 and shown in Table 9.6 (which also includes differences from TIA/EIA 568B).

Table 9.6 gives the allowable attenuation (from ISO 11801 and TIA/EIA 568B) of each building block, i.e. the fiber, the connector and the splice. The total attenuation allowed in each part of the link as also specified, and the designer must ensure that the total sum of attenuation figures does not exceed that allowed for that type of link. The links are defined as the horizontal cabling (up to 100 metres), the building backbone (up to 500 metres) and the campus backbone (up to 1500 metres). The specification has attenuation figures for multimode and single mode (note there is no differentiation between 50/125 and 62.5.125) and for operation at 850 and 1300 nm. The data is shown in Table 9.7.

Table 9.7 *ISO 11801 1st edition, channel attenuation allowance*

Cabling	Link length	Channel attenuation dB max			
		Multimode		Single mode	
Subsystem	Max	850 nm	1300 nm	1310 nm	1550nm
Horizontal	100 m	2.5	2.2	2.2	2.2
Building backbone	500 m	3.9	2.6	2.7	2.7
Campus backbone	1500 m	7.4	3.6	3.6	3.6

ISO 11801 2nd edition, however, changes the definition of these links into OF-300, OF-500 and OF-2000. The attenuation of these links is given in Table 9.8.

It is then incumbent upon the designer and installer to design the optical channel so that it meets the ISO 11801 design rules. The installer,

Table 9.8 *ISO 11801 2nd edition, channel attenuation allowance*

Cabling	Link length	Channel attenuation dB max			
		Multimode		Single mode	
Subsystem	Max	850 nm	1300 nm	1310 nm	1550nm
OF-300	300 m	2.55	1.95	1.8	1.8
OF-500	500 m	3.25	2.25	2.0	2.0
OF-2000	2000 m	8.5	4.5	3.5	3.5

Table 9.9(a) *Optical LAN operation, by power budget*

Network application	Optical power budget ISO 11801 2nd edition in normal type, *TIA/EIA-568-B.1 in italics*					
	Multimode		Single mode			
	850 nm		1300 nm		1310 nm	
10BASE-FL,FB	12.5 (6.8)	*12.5 (7.8)*				
100BASE-FX			11.0 (6.0)	*11.0 (6.3)*		
1000BASE-SX	2.6	*3.2 (3.9)*				
1000BASE-LX			2.35	*4.0 (3.5)*	5.0	*4.7*
10 000BASE-SX	draft					
10 000BASE-LX			draft		draft	
Token ring 4,16	13.0 (8.0)	*13.0 (8.3)*				
ATM 155	7.2		10.0 (5.3)		7.0	*7.0 to 12.0*
ATM 622	4.0		6.0 (2.0)		7.0	*7.0 to 12.0*
Fiber channel 1062	4.0				6.0	*6.0 to 14.0*
FDDI			11.0 (6.0)	*11.0 (6.3)*	10.0	
The figures in parentheses are for 50/125 performance						

having installed the optical cable plant, must then test it to ensure that measured optical attenuation equals or falls below the design figure. Only then can the operation of optical LANs be predicted and guaranteed over that cable plant. If the optical rules of ISO 11801 have been obeyed then the optical LANs detailed in Tables 9.9(a), (b) and (c) will operate over the distances listed.

Optical fiber highway design 183

Table 9.9(b) *Optical LAN operation, by fiber type and channel type*

Network application	ISO 11801 channel support (summary only)						
	OM1		OM2		OM3		OS1
	850 nm	1300 nm	850 nm	1300 nm	850 nm	1300 nm	1310 nm
10BASE-FL,FB	OF 2000		OF 2000		OF 2000		
100BASE-FX		OF 2000		OF 2000		OF 2000	
100BASE-SX**	OF 300		OF 300		OF 300		
1000BASE-SX	OF* 300		OF 500		OF 500		
1000BASE-LX		OF 500		OF 500		OF 500	OF 2000
10GBASE-LX4/LW4		OF 300		OF 300		nys***	OF 2000
10GBASE-ER/EW							OF 2000 (1550 nm)
10GBASE-SR/SW					OF 300		
10GBASE-LR/LW							OF 2000
Token ring 4,16	OF 2000		OF 2000		OF 2000		
ATM 155	OF 500	OF 2000	OF 500	OF 2000	OF 500	OF 2000	OF 2000
ATM 622	OF 500	OF 500	OF 500	OF 500	OF 500	OF 500	OF 2000
Fiber channel 1062	OF 500		OF 500		OF 500		OF 2000
FDDI		OF 2000		OF 2000		OF 2000	OF 2000 smf-pmd

*Note that IEEE 802.3z quotes 220 m for 160 MHz.km fiber and 275 m for 200 MHz.km fiber
**100BASE-SX is described in TIA/EIA-785
***nys: not yet specified

Table 9.9(c) *Optical LAN operation, by maximum supportable distance*

Network application	Maximum supportable distance ISO 11801 2nd edition in normal type, *TIA/EIA-568-B.1 in italics*					
	Multi mode				Single mode	
	850 nm		1300 nm		1310 nm	
10BASE-FL,FB	2000 (1514)	*2000 (2000)*				
100BASE-FX			2000 (2000)	*2000 (2000)*		
1000BASE-SX	275 (550)	*220 (550)*				
1000BASE-LX			550 (550)	*550 (550)*	2000	*5000*
100BASE-SX*	300 (300)					
Token ring 4,16	2000 (1571)	*2000 (2000)*				
ATM 155	1000 (1000)	*1000 (1000)*	2000 (2000)	*2000 (2000)*	2000	*15 000*
ATM 622	300 (300)	*300 (300)*	500 (330)	*500 (500)*	2000	*15 000*
Fiber channel 1062	300 (500)	*300 (500)*			2000	*10 000*
FDDI			2000 (2000)	*2000 (2000)*	2000	*40 000*

The figures in parentheses are for 50/125 performance, otherwise 62.5/125 fiber
*100BASE-SX is described in TIA/EIA-785

Transmission wavelength and the optical power budget

It will be noted that in the above example the first window was referred to rather than the specific wavelength of 850 nm. The reason for this is that much of the equipment that operates in the first window does not actually operate at 850 nm. More frequently it is found that LED-based

Table 9.10 *First window wavelength correction*

Measurement wavelength nm								Figures represent additional cabling attenuation (dB/km)
890	2.5	2.1	1.6	1.2	0.8	0.4		
870	2.1	1.7	1.2	0.8	0.4		−0.4	
850	1.7	1.3	0.8	0.4		−0.4	−0.8	
830	1.3	0.9	0.4		−0.4	−0.9	−1.2	
810	0.9	0.5		−0.4	−0.8	−1.2	−1.6	
790	0.4		−0.5	−0.9	−1.3	−1.7	−2.1	
770		−0.4	−0.9	−1.3	−1.7	−2.1	−2.5	
	770	790	810	830	850	870	890	Operating wavelength nm

sources operate on a broad spectrum around 820 nm or even 780 nm. It should be realized that the attenuation of the optical fiber will be significantly higher at these wavelengths than the measured value at 850 nm. The Rayleigh scattering losses dominate at these lower wavelengths and Table 9.10 indicates the level of attenuation change around that central wavelength.

The second window obviously has a direct and immediate impact upon the cable attenuation but the optical power budget calculation can be undertaken in the same manner (using the figures for power coupled and detected in the second window). Rayleigh scattering has far less influence in this window and it is not as necessary for the wavelength variations to be taken into account.

Component losses such as fusion splices and demountable connectors do not vary drastically between first and second windows; however, microbending within components can have significant effects at 1550 nm. Few systems in data and military communications yet operate in the third window and it will not be considered in detail.

Bandwidth requirements

The calculations of optical power budget and its implications for fiber geometry and span complexity show that in most cases there is no ideal fiber for a particular transmission system. A range of fibers could be used and choosing the most appropriate one is not necessarily a straightforward decision. The decision-making process in the selection of the right optical fiber type for a particular campus, backbone or high-connectivity highway must start, obviously, with an assessment of whether the proposed transmission equipment will operate. Where there are alternatives other factors must be considered and the first of these is the operational bandwidth of the highway.

It is easy to see that most spans will operate with maximum margin using optical fibers with the largest optical core diameters and highest numerical apertures. This is certainly true provided that the optical fiber attenuation does not become too great. There are two arguments against using the maximum diameter and NA fibers. The first is cost: as was indicated in Chapters 2 and 3 the cost of producing optical fiber increases with cladding diameter and dopant content. The second, and by far the most important, factor is that of bandwidth.

As will be recalled from earlier chapters the bandwidths of optical fibers increase as core diameter and numerical aperture decrease. This is a key factor in determining the geometry to be chosen in large-scale campus and backbone environments.

With the exception of high-connectivity highways (which are discussed below) the only contenders are the professional grade data communications and telecommunications fiber geometries:

single mode   8/125 μm, 0.11 NA
multimode   50/125 μm, 0.20 NA
               62.5/125 μm, 0.275 NA

This section reviews these geometries and discusses their application to the optical cabling infrastructure.

62.5/125 μm optical fiber

The 62.5/125 μm geometry was first proposed by AT&T in the USA as being a medium bandwidth optical fiber which featured a good level of light acceptance. The established 50/125 μm design, which is still common both in Europe and Japan, offered higher bandwidth at the expense of light acceptance. Despite the fact that the vast majority of transmission systems could operate satisfactorily in moderately complex configurations using 50/125 μm, it was thought that 62.5/125 μm may have long-term advantages.

The 62.5/125 μm design is used widely in the USA and elsewhere despite its cost penalty over 50/125 μm and can be supplied by a variety of manufacturers.

The bandwidth of TIA/EIA 568A and FDDI specification 62.5/125 μm optical fiber is as shown below:

	850 nm	1300 nm
Bandwidth (min)	160 MHz.km	500 MHz.km

The bandwidth of ISO 11801 and EN 50173 62.5/125 μm optical fiber is:

	850 nm	1300 nm
Bandwidth (min)	200 MHz.km	500 MHz.km

The implications of these bandwidths upon campus and backbone environments are discussed below.

50/125 μm: the ultimate multimode optical fiber?

In Europe and Japan this geometry dominated non-telecommunications applications before the arrival of 62.5/125, and has been used to service complex cabling infrastructures.

The optical attenuation of 50/125 μm and 62.5/125 μm geometries is broadly similar and typical values are shown below:

Attenuation (min)	850 nm	1300 nm
50/125 μm	3.0 dB/km	1.0 dB/km
62.5/125 μm	3.75 dB/km	1.75 dB/km

As the vast majority of campus and backbone applications do not feature active link lengths of greater than 1 kilometre the differential attenuation has little meaning. Much more relevant are the bandwidths available in the 50/125 μm geometry. The bandwidth of general specification 50/125 μm optical fiber is as shown below:

	850 nm	1300 nm
Bandwidth (min)	400 MHz.km	800 MHz.km

It is possible to purchase enhanced bandwidth fibers giving bandwidths of 1000 MHz.km (850 nm) and 1500 MHz.km (1300 nm). Process limitations limit the same level of availability of bandwidth range for high dopant content fibers such as 62.5/125 μm designs. The OM3 optical fiber of ISO 11801 2nd edition is 50/125 with a 500 MHz.km overfilled bandwidth at both 850 and 1300 nm, but with a 2000 MHz.km laser launch bandwidth at 850 nm.

8/125 μm: the ultimate fiber – the cheapest and the best?

A great debate has taken place over the years with reference to the 'correct' choice of multimode optical fiber. For campus and backbone environments this argument has rationalized itself into 62.5/125 μm versus 50/125 μm. There are two reasons for this.

The first reason is that the bandwidths of these multimode fiber designs have been adequate for the vast majority of current transmission applications. The longest typical active span within a campus style infrastructure is 2 kilometres. In Chapter 2 it was stated that bandwidth tends to be linear over such distances and therefore the first window bandwidths will be a minimum of 80 MHz (62.5/125 µm) and 200 MHz (50/125 µm). This normally equates to 40 and 100 megabits/s respectively. These data rates are adequate for applications such as Fast Ethernet (100 Mb/s) and ATM up to 155 Mb/s. As a result there was little desire to progress to higher-bandwidth windows (1300 nm) or to higher-bandwidth geometries (single mode).

Second, the costs of integrating a single mode system have been too high in comparison to the equivalent multimode system. Single mode optical fiber (8/125 µm) is by far the cheapest geometry. Reference to Chapter 3 will remind the reader that the step index, low NA (low dopant content) nature of the single mode design produces a product at approximately one-third of the cost of 50/125 µm formats (and one-fifth of the cost of 62.5/125 µm). Unfortunately the cost of injecting power into the small core has historically been quite high (due perhaps to the telecommunications *cache* of the products involved) and the cost of the transmission equipment has therefore swamped the cost benefits of the cheaper fiber type. Obviously in long-distance high-bandwidth systems there are savings overall due to the reduced need for repeaters/regenerators, but these are not reflected in the campus and backbone environments.

However, these factors are coming under attack. The installation of cabling infrastructures, which are expected to exhibit extended operational life, puts considerable pressure on the ability to predict service requirement. The introduction of gigabit and ten gigabit standards have made it necessary to migrate to single mode.

This uncertain science of prediction would not, on its own, impact the uptake of single mode optical fiber. However, it is being combined with another much more important factor. Around the developed world optical fiber is being viewed as a means of providing communication to the domestic subscriber, i.e. the home. These far-sighted projects are not being undertaken out of altruism. Rather they are aimed to allow the telephone carriers (PTT) to offer a wide range of high-quality communications to the home which will include interactive products such as home shopping, home banking as well as an enormous variety of video networking services.

These wideband services will necessitate the use of single mode technology and it is obvious that these services must be offered commercially. This will drive down the cost of single mode sources and detectors and will also initiate the development of new sources capable of launching adequate powers into the 8 µm core. Devices, such as 1300 nm single mode VCSELs (vertical cavity surface emitting lasers) will be of a lower output than the long-range telecommunications products and will be suitable for the local network.

Based upon these trends it is highly likely that the great debate between multimode geometries will continue for some time but that by 2005 much of the campus and building backbone infrastructures will be installed using single mode optical fibers. With the exception of certain special systems, it is probable that single mode will totally dominate the market by the end of the decade.

Fiber geometry choices within the highway design

The discussion of the bandwidths of the various optical fiber geometries is important in the design of the highway. The choice of optical fiber geometry must reflect the needs of the predicted services to be provided on the highway. As was indicated above the obvious choice would be to use single mode throughout but this would obviously limit the initial operation of the network using existing multimode transmission equipment.

It is therefore not unrealistic to install large campus networks within cabling infrastructures containing two or more fiber geometries — multimode optical fiber is included to meet current and medium-term needs while single mode elements are included to guarantee the operation of fast services not yet required. The single mode content is achieved at minimal material cost and is rarely terminated in the initial configuration. The choice for the multimode content must be made according to the above technical arguments and occasionally the inclusion of both 50/125 µm and 62.5/125 µm fibers is agreed.

While it is technically acceptable to mix multimode geometries within a single site (and even a single cable construction) great care should be taken to ensure that interconnection of the different geometries cannot take place accidentally, thereby rendering the network both inoperative and difficult to repair quickly.

'To-the-desk' applications and fiber geometry

The horizontal or 'to-the-desk' intra-building infrastructure typically contains much shorter links than the campus environment. For this reason it is feasible to adopt lower bandwidth fiber designs

These factors combine to suggest the use of a 62.5 µm geometry to maximize the optical power budget whilst not limiting the bandwidth of the installed cabling.

There are various factors which may further influence the choice of geometries adopted. First, if the intra-building cabling scheme is part of a larger inter-building cabling infrastructure, then the services offered to the various nodes within the buildings may suggest the use of the same geometries as in the inter-building cabling. Second, the issue of service expansion or modification must be addressed, not from the technical aspect but from the commercial viewpoint.

Centralized optical architecture (COA) means running one fiber directly from the work area back to the central equipment room, thus doing away with copper to optical electronics, horizontal cross-connects and floor distributors (i.e. patchpanels on each floor). For the larger installation, where the average distance from the desk or work area to the central equipment room is more than 100 metres, COA becomes a much more commercially attractive option. The introduction of cheaper optical transmission systems, such as the 100 Mb/s Ethernet 100BASE-SX, will also accelerate this process.

COA was first recognized with the standard known as TSB 72, i.e. Telecommunications Service Bulletin number 72 of the TIA/EIA 568A Commercial Premises Cabling Standard. TSB 72 acknowledges that it is unnecessary to impose 100 metre horizontal cabling rules, with a compulsory horizontal cross-connect, when dealing with optical fiber. You can just run it from the desk directly back to the computer room. A distance of 300 metres was allowed for, presumably to match the forthcoming gigabit Ethernet over multimode fiber standard. The gigabit Ethernet standard, IEEE 802.3z (1000BASE-SX), subsequently imposed a limit of just 220 metres as the guaranteed minimum operating distance, but the 300 metre limit has remained in the next edition of TIA/EIA 568B.

The second edition of ISO 11801 has taken the idea of centralized optical architecture to its logical conclusion, however, and allows a direct fiber link from the wall outlet of up to 2000 metres on multimode fiber or single mode fiber. Figure 9.9 shows the concept of COA from the ISO 11801 2nd edition model.

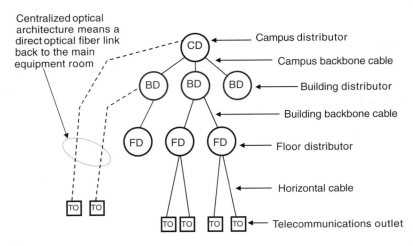

Figure 9.9 *Centralized optical architecture (ISO 11801 2nd edition)*

High connectivity and extreme environment cabling

In the campus and backbone applications it has been assumed that the installed cabling will experience a relatively benign environment. It is not normal for data communications cabling to be subject to temperatures higher than 70°C or lower than −20°C. Neither is it typical for the optical fiber to be put under high atmospheric pressure. However, these extreme environments do exist in special circumstances and if cabling is to operate under such conditions the choice of fiber geometry can be as important as the design of the cables and connectors to be used.

In general the impact of extreme conditions is seen in increased levels of attenuation over relatively short distances. The applications in which such conditions are applied frequently incorporate high connectivity levels which may add to the attenuation problems encountered.

Extreme temperature and pressure

High-temperature operation reduces the optical output power from LED sources but, if not controlled, increases the output from laser sources. Low temperatures reverse these effects and before making any assessment of the fiber requirements it is necessary to establish the worst case operational optical power budget.

With regard to optical fiber the effects of low temperature and high pressure are similar. Microbending at the CCI (core–cladding interface) takes place and can significantly increase the attenuation, even over a short distance.

The ability of the optical fiber to guide the light transmitted is related to the NA. To minimize the effect of microbending it is necessary to adopt high NA geometries. This leads to lower-bandwidth cabling solutions which over short distances are not normally a problem (obviously where high bandwidths are required it is necessary to prevent the effects of the temperature or pressure from reaching the optical fiber itself).

The adoption of high NA designs for their guiding properties at low temperature (or under high pressure) also increases the light acceptance from LED sources at elevated temperatures, thereby maximizing the opportunity for the cabling to meet the worst case optical power budgets.

High connectivity

The need for high-connectivity cabling has direct implications for the fiber geometry. The need for operation over significant numbers of demountable connectors can produce attenuation levels which exceed those of long-range systems. As a result large core and high NA geometries may be adopted at the expense of bandwidth.

The use of 100/140 μm designs represents the first step in response to these needs and the 0.29 NA gives a desirable degree of light acceptance and microbend resistance. To achieve improvements beyond 100/140 μm the range of 200 μm core diameter designs can be used. There are, however, a variety of designs, some of which are detailed below:

- 200/230 μm: 200/250 μm: 200/280 μm: 200/300 μm;
- 0.20 NA: 0.30 NA: 0.40 NA: 0.45 NA.

Summary

The design of an optical fiber highway has to address a number of issues relating to the choice of the transmission medium, its route and the philosophy of repair and maintenance.

Once installed the design has to be capable of expansion and reconfiguration of the services offered together with evolution within the technology including changes in transmission wavelength.

Chapters 10–16 discuss the methods to be used to move from a paper design, or operational requirement, to a well-specified, well-installed cabling infrastructure.

Reference

1 Elliott, B. (2000), *Cable Engineering for Local Area Networks*, Cambridge: Woodhead Publishing.

10 Component choice

Introduction

The outline design of an optical fiber cabling infrastructure can be produced without reference to the specific components to be used. As discussed in Chapter 9 the optical fiber geometry must be chosen to meet the initial and future requirements of the transmitted services but the other aspects of the design, such as the termination enclosures, cables, connectors and jointing techniques, have been treated purely in terms of their optical performance.

The choice of components is not trivial and should consider a number of installation and operational issues. These issues and the recommendations that result are detailed in the following sections.

The components chosen for a particular cabling task must first be capable of surviving the process of installation (which may be regarded as a combination of physical and environmental conditions for that specific installation). Once installed the components must be able to provide the desired level of optical performance over the predicted lifetime of the cabling whilst enduring the assault of relevant mechanical and climatic conditions. This chapter seeks to give general guidance allowing the reader to determine the final choice based upon a rational approach.

Fiber optic cable and cable assemblies

Cable is purchased in various forms:

- *Fixed cable.* Defined in Chapter 7 as cable which once installed cannot be easily replaced and is contained within termination enclosures at either end.

- *Pigtailed cable assemblies*. Terminated cables with demountable connectors at one end only. They may be manufactured in a number of formats depending upon their application including:
  - SCOF elements for subsequent jointing to fixed cable elements within termination enclosures configured as patch panels. The secondary coated element has no integral strength member and cannot be used external to the termination enclosure without risk of damage to the termination;
  - SROFC elements for subsequent jointing to fixed cable elements within termination enclosures. The strength member within the SROFC is tied to the chassis of the termination enclosure and the cable normally passes through a gland or grommet in the wall or panel of the enclosure. The single ruggedized elements can be handled directly and can be directly connected to transmission equipment;
  - SROFC pigtailed cable looms feature either multiple SROFC elements within one overall sheath or within a loose cableform. These are used in the same manner as the SROFC elements above; however, it is normal for these assemblies to form part of a distribution network from a termination enclosure to a remote equipment rack.
- *Jumper cable assemblies*. Terminated cables with demountable connectors at both ends. These form the most basic method for interconnecting transmission equipment, connecting transmission equipment to termination enclosures or, in the case of high-connectivity highways, interconnecting the subnodes at bulkheads etc. As in the case of pigtailed cable assemblies the jumper cable assemblies may be manufactured in a variety of formats including the following:
  - *simplex*: a single-element SROFC terminated with the desired demountable connectors at either end;
  - *duplex*: a twin-element cable comprising two SROFC units terminated with single ferrule or dual ferrule connectors;
  - *looms*: multiple SROFC elements within one overall sheath or within a loose cable form. These may be terminated with multi-ferrule connectors or with individual single ferrule connectors.
- *Patch cable assemblies*. Terminated cables with demountable connectors at both ends with the express purpose of connecting between the various ports of patching facilities. As a result these cable assemblies normally have the same style of connectors at both ends and are simplex SROFC in format to maximize flexibility.

It will be noticed that the pigtailed, jumper and patch cable assemblies are based upon SCOF, and where a true cable construction is adopted, then this is in the form of SROFC. No loose tube designs are discussed for these types of cabling components since the token presence (or total absence) of an effective strength member (for attachment to the demountable connector) is not acceptable for cables which may be expected to receive regular and uncompromising handling. In any case the cost savings attributable to the use of loose tube constructions on short lengths are minimal and cannot be balanced in any sensible way against the potential risks of damage.

This section reviews the choices for these various fiber optic cables and assemblies in order.

Fixed fiber optic cables

There are a great variety of cable constructions and material choices open to the installer; however, it should be realized that not all variants are instantly available. As in any industry the customer can purchase a product to an exact design and specification provided that sufficient quantity is required, sufficient payment is made and sufficient time is allotted. As a result the standard products available offer basic constructions meeting a generic specification and deviation from these basic designs often results in extended delivery, minimum order quantities and premium pricing.

The most effective method of choosing a suitable fixed cable design is to review its installation and operating environment. Chapter 7 has already discussed constructions in detail but this guide may be used as a reminder.

Environment	*Recommended construction*
Direct burial	Armoured: wire armour is normal, but corrugated steel tape is also acceptable, unless metal-free requirement exists.
	Polyethylene sheathing materials are normal due to their abrasion resistant properties.
	Central strength members are included to act as a means of installation. Metal is normal unless a metal-free requirement exists.
	Moisture resistance is achieved by the inclusion of foil or tape-based barriers beneath the sheathing materials. If it is possible for moisture to enter the cable from the ends (perhaps from drawpits) then a gel-fill may be desirable.
	Loose tube constructions are desirable to prevent installation stresses being applied to the optical fibers.
Catenary installation	Any catenary (aerial) installation is a potential lightning hazard and cables designed for this purpose are normally metal-free, using insulating central (or wrapped) strength members and non-metal moisture barriers.

Duct or other   As for direct burial, however, the requirement for armour can be relaxed or removed.

Loose tube constructions are desirable to prevent installation stresses being applied to the optical fibers.

Internal   Internal applications are generally not as physically demanding for the cable constructions either at the installation stage or during fixing and operation. Typical methods of fixing are on cable trays, in conduit or trunking. The degree of flexibility needed to wind along cable runs within buildings is generally greater than that in external ducts or traywork and smaller cables are necessary. This favours the use of tighter constructions incorporating SCOF elements.

The strength members are generally yarn based and also serve as impact-resistant layers.

The sheath materials have historically been PVC based but fears with regard to toxic gas generation during combustion have led to the use of a variety of low fire hazard and zero halogen materials.

To meet the requirements of a particular cable route it may be necessary to pass through both internal and external environments and in such cases care must be exercised in choosing the design of cable. The normal sheathing materials for external cables are polyethylene based which although not toxic are not always accepted for internal use due to their flammability.

The options therefore are either to use a single cable design featuring a material with low fire hazard both internally and externally or to joint between two cable designs at the entrance to the building.

The first option is not acceptable for indoor-only rated cables because the low fire hazard (LFH) materials tend to be rather easily abraded and as such are not suitable for direct burial or duct installation. Another reason for not using LFH sheaths in the external environment is that they are hygroscopic (moisture absorbing) and can act, in the most extreme cases, as a conduction path under high electrical potentials such as lightning discharge.

The second option may be undesirable due to the additional losses generated at the joints plus the expense of the extra labour required to make the transition joints.

For campus installations where cables will be expected to be run within and then in between buildings, then universal grade cables, i.e. cables with a weather/water resistant *and* a fire resistant sheath, are often the most cost-effective option.

The impact of moisture upon optical fiber

The effect of moisture was mentioned above with regard to absorption within the LFH sheath materials. It was also discussed in Chapter 7 with regard to direct ingress to the cable construction via sheath damage or by capillary action along the optical fibers from the cable ends. At this point it is worthwhile examining the actual effect of moisture on the optical fiber.

Optical fiber can be affected by moisture in two ways. First, optical fiber under stress may fail by the propagation of cracks. The rate of propagation is accelerated by environmental factors of which humidity is a key element. It is good practice to install loose tube constructions in environments where moisture can be a potential problem (in water-filled ducts, for example). The loose construction allows the optical fibers to lie within the cable under little or no stress and the presence of moisture is less catastrophic. The second aspect is the large-scale penetration of water or other liquids into the cable construction in such a way that they might freeze. In particular water expands when it freezes and in a confined space it elongates, thereby putting axial tensile and compressive radial loads upon the optical fiber which may subsequently fail.

The key design feature for a cable which may be subjected to rugged installation practices in wet environments is the prevention of moisture ingress. A strong polyethylene sheath is a good foundation; however, if that becomes damaged, then there is a definite need for a moisture barrier. There are various styles. The most common is the polyethylene/ aluminium laminate wrap which seeks to prevent the passage of moisture further into the cable. An alternative is to redirect the moisture flow away from the optical fibers. This is sometimes achieved by the incorporation of two sheaths separated by a yarn-based layer. This layer acts as both an impact-resistant barrier around the cable and as a moisture conductor that absorbs the moisture by capillary action. The total amount of moisture which can be absorbed is limited by the actual free volume within the yarn layer and subsequent freezing can have no effect upon the optical fibers beneath the inner sheath (see Figure 10.1).

As an added insurance it is possible to prevent moisture travelling along the optical fibers (from drawpits etc.) by the inclusion of a gel-fill. The gel fills up the interstitial spaces within the cable and prevents water ingress very effectively. It can be very messy and time consuming to terminate gel-filled cables, however. A modern alternative is to put

198 Fiber Optic Cabling

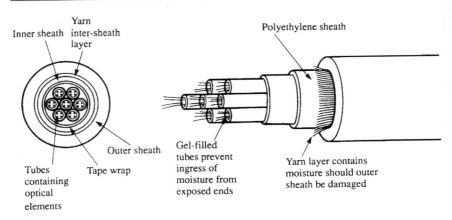

Figure 10.1 Non-metallic moisture barriers

water-swellable threads and tapes within the cable. These threads and tapes will swell up if they are in contact with water and can thus prevent the spread of water along a cable.

Identification of optical fibers

It is highly desirable that each optical fiber within a cable construction should be uniquely identifiable. For a large fiber count loose tube cable this may be achieved by the colouring of the PCOF or SCOF elements within each tube (or slot) and with each tube being uniquely identified by virtue of colour or position within the overall construction.

The human eye can only differentiate about 12 colours (in a cable installation environment) and so bundles of 12 primary coated fibers are wrapped in a colour coded thread to further differentiate and identify each bundle.

It should not be forgotten that cables may contain more than one fiber design. In such cases the cable construction should be arranged in a fashion that allows easy recognition and handling of the separate designs.

Cable identification

For many years users in all application areas have asked for optical fibers to be manufactured and sheathed with a material of a standardized and exceptional colour in order that fiber optic cables might be immediately recognizable among other types of carrier. This has not been achieved.

First, sheath materials which are under attack from ultraviolet radiation tend to suffer from deterioration and breakdown if they are not black in colour. Second, there are few, if any, colours which could be adopted as a standard which are not already used in some copper-cabling applications.

As a result it is possible to purchase internal grade cables in almost any colour that the user desires – providing that the quantity is cost effective. However, most external grade cables continue to be made in black only.

Pigtailed, jumper and patch cable assemblies

Where the cabling infrastructure includes optical fiber of a single design the choice of colour for the SROFC assemblies is arbitrary since little confusion can arise. However, where multiple geometries are used it is vital to differentiate between the various ruggedized cable accessories.

Cable colours should be selected for each fiber geometry and strictly adhered to throughout the initial installation and for all modifications and reconfigurations undertaken during the life of the highway.

Connectors

Equipment and system connectors

The demountable connectors to be used within cabling infrastructures can be categorized as equipment connectors and system connectors.

The equipment connector is that which connects directly to the transmission equipment. For a given highway there may be more than one equipment connector depending upon the variety of services operating or even due to differing manufacturers.

Equipment connectors are normally purchased as preterminated cable assemblies such as jumper cable assemblies or SROFC pigtailed cable assemblies (which will be jointed to fixed cables).

Transmission equipment is supplied in three basic formats:

- *Receptacle based.* The source and/or detector components are mounted in receptacles or sockets which are fitted directly to the equipment chassis. The receptacles are constructed to allow an equipment connector of the same generic design to be attached, thereby enabling an adequate level of light injection into the optical fiber.
- *External pigtail based.* The source and/or detector are fitted with pigtails which emerge from the transmission equipment, enabling direct connection to an equipment connector in an adaptor fitted to an external chassis or similar.
- *Internal pigtail based.* The source and/or detector are fitted with pigtails which are connected to adaptors on the chassis of the transmission equipment.

The system connector is that connector which is adopted for all termination enclosures and facilities. This is not necessarily of the same generic design as the equipment connector. Instead the system connector should reflect current trends in standardization and miniaturization and should be chosen for reason of its performance, packing density and reliability.

Although the cabling standards still recommend the use of the SC duplex optical connector, it has to be recognized that optical transmission equipment will come with an almost random selection of SC, ST, MT-RJ, SG, LC, FC-PC or Fiber Jack connectors.

A decision needs to be made to select connectors in the cross-connect environment, i.e. at the patch panels, which represents the most likely (if known) manifestation of the equipment connectors coupled with a desire for high-density, high-reliability and low cost patching. In fiber-to-the-desk installations, it is possible that a totally different connector be selected for the telecommunications outlet role.

It is a firm recommendation that each demountable joint within a cabling design should be produced using single-source components. This will ensure compliance with the optical specification and provide the necessary contractual safeguards.

Splice components

Demountable connectors come under particular scrutiny because they represent a variable attenuation component within the optical loss budget. The choice of splicing techniques and components is no less important, particularly since a failure of these joints is potentially more difficult for the inexperienced user to locate and repair.

Chapter 6 discussed the cost implications of the different splicing techniques and it was shown that the mechanical splice joint offered benefits for the irregular user whereas the more committed installer would probably adopt the fusion method on the grounds of cost, simplicity and throughput.

Independent of the technique the components used must fulfil the requirements of stable optical performance over an extended time scale.

Fusion splice joints are primarily process based and, providing that a good protection sleeve is used, the final performance is largely dependent upon the quality of the equipment used. Mechanical splices, however, are highly dependent upon the components used and as a result the final choice should consider the following issues:

- *Mechanical strength of the joint.* This is centred around the tensile strength of the fiber bond (whether mechanical or adhesive in nature) and the variations of that strength under vibration, humidity and thermal cycling.

- *Stability of any index matching fluids or gels present.*
- *Strength of any cable strain relief present.*

Termination enclosures

Because they are not optical fiber components the termination enclosures are frequently ignored and not treated as part of the installation. Nevertheless, the use of the wrong design of termination enclosure can be a major influence upon the reliability and operational lifetime of the overall fiber optic cabling scheme.

The term 'termination enclosure' covers virtually all the types of housing which might be used in the installation of the cabling. Their purpose is to provide safe storage for, and access to, individual optical fibers within the cables used. It must be noted that single mode fiber is much more sensitive to bending compared to multimode fiber being used at 850 nm, thus techniques used for fiber management that may only just be suitable at 850 nm will not be possible when using single mode fiber.

Typical applications are:

- Jointing of fixed cables within extended routes. The majority of fixed cable designs are manufactured in long lengths but are generally purchased in 1.1 km or 2.2 km lengths. If routes exceed the purchased lengths then it may be necessary to joint cables at suitable locations.
- Change of fixed cable type. For instance, the fixed cable may have to be changed at the entry point to a node where external cable designs must be converted to those suitable for internal application.
- Change of cabling format and/or capacity.
- Termination of fixed cable in a pigtailed format.
- Termination of fixed cable in a patch panel format.
- Housing of other components such as:
  - splitters, branching devices;
  - optical switches;
  - active devices and transmission equipment.

Figure 10.2 shows all the above applications.

External features

Termination enclosures are manufactured from a range of materials, both metallic and non-metallic, and are designed for mounting against walls, within cabinets, under floors, upon poles, in drawpits and manholes, and even in direct burial conditions.

Consideration should be given to the relevant environmental factors including temperature, humidity and vibration together with the less

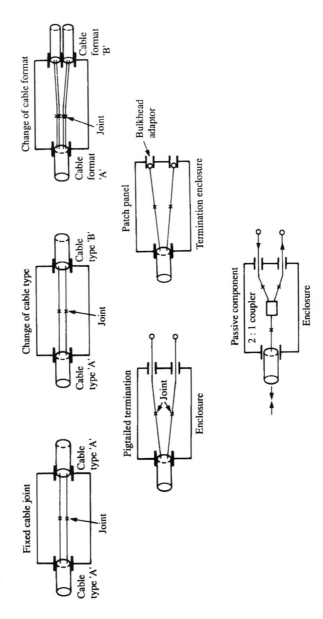

Figure 10.2 Termination enclosure schematics

obvious conditions such as ambient lighting, fluid contamination, mould growth and so on.

Obviously a termination enclosure is, in effect, a convenient point of access to the optical fibers within the cabling. Nevertheless the enclosure represents a potential source of unreliability (due to the probability of user interference) and also a possible area of insecurity or sabotage (should this be an issue). For this reason the termination enclosures are frequently provided with some type of security feature to prevent unauthorized access.

In all cases the termination enclosure must provide suitable strain relief to each cable entering the enclosure and where necessary must maintain the environmental features of the fixed cables used (the cabling generally enters the enclosure through a gland which must not affect the degree of environmental protection provided by the unpenetrated enclosure). Where relevant the termination enclosure should include the necessary fittings to provide earth bonding to any conduction path within the fixed cable.

Internal features

The necessary internal features of a termination enclosure relate to fiber management. In general the complexity of the enclosure is limited by the ability of the optical fibers to be installed, modified or repaired. For instance, a 19-inch subrack enclosure can support up to approximately 48 demountable connectors in line across the front panel; however, the difficulties in fiber management within the enclosure often limits this number to 24 or even 12 for a 1U (i.e. 44 mm) subrack.

There are various methods used to manage optical fiber and any joints within the enclosure. The key factor to be addressed is whether or not it is straightforward to access a given element, remove it, undertake jointing or repair and finally replace it without disrupting the communication on neighbouring elements. It is also desirable for the fiber management methods to allow easy identification of the individual elements. This may be achieved by the use of coloured fibers (or sleeves), alphanumeric labelling or physical routeing.

Access to termination enclosures

An area which frequently does not receive sufficient attention is that of access to the enclosures themselves. At the time of initial installation a minimum of 5 metres of fixed cable should be left at the location of each enclosure. Consideration should then be given to the future method of gaining access to the termination enclosure and for its removal.

Summary

The various cable configurations, the demountable connectors, the permanent and temporary joint techniques and components are responsible for the optical performance of the installed cabling. The termination enclosures, despite not being responsible for the initial performance of the cabling, can, if not chosen correctly, impact its operational aspects including reliability and repairability.

11 Specification definition

Introduction

Thus far this text has concentrated on the theory of optical fiber, its connection techniques, the design of the cabling infrastructure and, in Chapter 10, the choice of components to be used within that infrastructure.

However, the installation of any fiber optic cabling scheme is not complete until it has been tested, commissioned, documented and handed over to the user. Moreover the cabling must be proved to be capable of providing the desired services at the desired locations (or nodes). These issues have little to do with the technical capabilities of the optical medium and are basically contractual in nature.

The vast majority of problems encountered during the installation tend to be contractual rather than technical. This is never more clearly highlighted than in the many invitation to tender documents that are received for which the operational requirement can be summarized as 'please supply a fiber optic backbone'. Such a vague request is in sharp contrast with the relevant sections on copper cabling which define the true requirement down to the last metre of cable and the last cable cleat. It is hardly surprising then that the final installation may not be all that it should be – but what should it be, since there was no firm specification?

This chapter provides the glue that holds the technical and the contractual issues together.

Technical ground rules

In many other parts of the world, single mode optical fiber has been in use since the early 1980s. The telecommunications market has developed faster and faster transmission technology to the point where 10 gigabits/second is not uncommon. In general these developments have not impacted the basic design of the optical fiber itself or the methods and components used to interconnect that optical fiber.

Before single mode technology was adopted the multimode fiber geometries such as 50/125 μm and 62.5/125 μm were well developed and more recent improvements in manufacturing processes and materials have allowed interconnection components to be produced which are probably as good as they are ever going to be.

The earlier chapters discussing fiber and connectors together with that on component choice suggest the existence of a mature market at the cabling level (with most of the developments being undertaken at the transmission equipment end of the market). The only trends in cabling development that defy this generalization are the connection mechanisms necessary for the widespread use of optical fiber in the office or the home. These are briefly discussed in Chapter 18. Nevertheless these are not technical issues and are instead dominated by commercial concerns.

As a result there is little that is unknown at the technical level with regard to what can be achieved and what cannot be achieved. There are only two reasons why a fiber optic cabling scheme will fail to operate: too much attenuation or too little bandwidth. Both parameters are well understood and provided that the scheme has been designed correctly and the correct components and techniques used it is rare for technical issue to be a source of dispute between installer and user.

However, this assumes that there is a clear understanding between the user and the installer of what is required, how that requirement is to be met, how it is to be proved that it has been met and, finally, what documentation and support services are necessary to ensure that the requirement will continue to be met for the predicted lifetime of the cabling infrastructure.

This is the purpose of the infrastructure specification – a document providing a contractual framework which, by addressing technical issues at the highest level, enables any subsequent problems to be resolved in a contractual manner without risk to the overall objective of providing the user with an operating optical fiber highway. In other words, once a specification has been defined, and agreed, the optical fiber issues can be regarded as having been resolved: the problems encountered will have little or nothing to do with the cabling medium and much more to do with who digs the holes (for example).

Operational requirement

The specification is the culmination of the design phase and suggests a common objective between the user and the installer and will consist of a number of separate documents. The operational requirement is the first of these documents and is followed by the design proposal and formal technical specification. The operational requirement is a statement of need and it may address the following issues.

Topology of the optical fiber highway

The purpose of the proposed cabling must be clearly defined. It is sensible to commence with a description of the topology of the infrastructure.

A list of the proposed node locations should be produced together with a straightforward and relevant coding system for these locations. The coding system may be in accordance with existing site convention but in the absence of this the system may be defined in the following way.

Campus and backbone cabling infrastructures

Buildings are most conveniently denoted by numeric codes, e.g. 01, 02, 03, . . . , 99.

Floors within buildings are most conveniently denoted by alphabetic codes, e.g. A, B, C, . . .

Individual nodes on floors within buildings are most conveniently denoted by numeric codes prefixed by the relevant floor code, e.g. A02, G06 etc.

A particular node may therefore be described as a combination of numeric and alphabetic codes such as 19A, which suggests the only node on Floor A within Building 19. Similarly Node 23B07 represents location 07 on Floor B within Building 23.

This convention for the coding of node locations is normally flexible enough to accommodate virtually any topology. Using a two-digit building code up to 100 buildings may be cabled, and each of them could support up to 26 floors, each of which could support up to 100 individual nodes. This is rarely exceeded and should this be the case the system can always be extended.

High-connectivity highways

High-connectivity highways take a large variety of forms but it is normal to denote the subnode locations by a numeric coding system. This enables a given cabling element to be specified by the use of the two subnode location codes to which it connects.

Coding systems for the nodes or subnodes such as those outlined above are invaluable at the planning stage. The codes allow the provision of a block schematic (see Figure 11.1). This block schematic is the basis for the operational requirement.

208 Fiber Optic Cabling

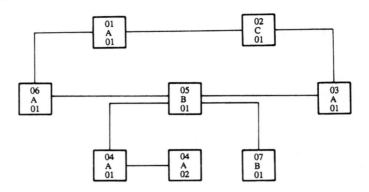

Figure 11.1 Block schematic

Block and cabling schematics

The block schematic merely shows the overall connectivity requirement and allows the production of a nodal matrix. The nodal matrix is a simple representation of all the internodal (or intersubnodal) connectivity. It ensures that all the proposed routes are included in the design. Each route

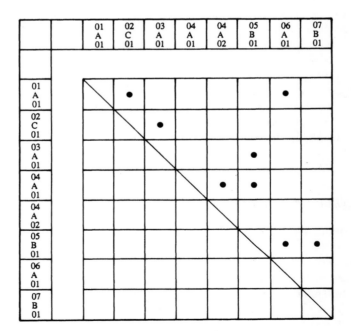

Figure 11.2 Nodal matrix

Table 11.1 *Route description*

Opening node	Closing node	Route description		Cable count/fiber count
01A01	02C01	External, direct burial	(750 m)	1/12
01A01	06A01	External, direct burial	(1250 m)	1/12
02C01	03A01	External, duct	(250 m)	1/12
03A01	05B01	External, duct	(600 m)	1/12
04A01	04A02	Internal, horizontal	(75 m)	1/4
04A01	05B01	External, catenary and aerial	(150 m)	2/6

must then be reviewed in terms of the cabling requirement. The nodal matrix shown in Figure 11.2 features an eight-node highway and Table 11.1 defines the nature of the route in terms of its cable requirement.

From the route description it is possible to define the cabling environment for each route. It may be that particular routes will feature both external and internal sections, which may suggest the use of further jointing points (to change from internal to external grade cable constructions). These jointing points will then be regarded as additional nodes.

In addition to the physical environment for the proposed cable it is necessary to define a route in terms of fiber capacity (fiber count and fiber design) and, where relevant, the need for full redundancy within the cabling. The latter involves the use of two or more separate routes between each pair of nodes.

This leads directly to the provision of a cabling schematic which details the number of fibers running between nodes together with a definition of the grouping of the fibers within cables on each route. A cabling schematic is shown in Figure 11.3.

Network configuration

As has already been stated it is pointless for a designer and an installer to produce a cabling infrastructure which cannot support the equipment needed to provide the services desired between the nodes.

The operational requirement must include details of the proposed usage of the highway, at least as it is to be initially operated. Using this information it is possible to produce overlays on the cabling schematic which show the various configurations of the highway providing the initial services and, where required, the configurations which would be adopted to provide new or revised services in the future.

For instance, the customer who desires all eight nodes to be serviced

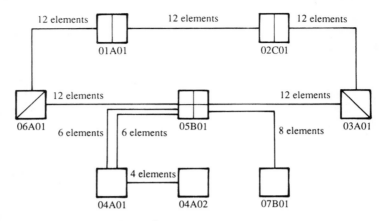

Figure 11.3 Cabling schematic

by Ethernet communications must be able to configure the installed cabling to meet that requirement.

Summary

Once installed the average customer will rapidly forget the fact that the services being communicated across the site, within the building or around the ship, are transmitted over optical fiber. The operational requirement is produced to ensure that the services required are transmitted in a manner that will allow expansion or evolution of those services without jeopardizing the investment already made in the cabling infrastructure.

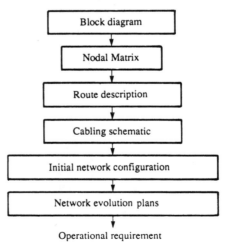

Figure 11.4 Operational requirement

Specification definition 211

An operational requirement enables the installer to produce a number of documents that are key to the agreement of the specification of the proposed infrastructure. A flowchart detailing these documents is shown in Figure 11.4.

Design proposal

In response to the operational requirement a design proposal may be produced either by the user or the installer or, as is more usual, by a combination of the two. It should address the following issues.

Component choice analysis

The operational requirement can, in theory at least, be produced without recourse to any knowledge of optical fiber theory, components or practices. The network configuration documents defined the services to be operated between the nodes. The flexibility of the infrastructure must be reviewed by considering the issue of pigtailed or patch panel termination enclosures.

The use of pigtailed or patch panel termination enclosures is determined by balancing the needs for flexibility against the additional attenuation produced. This decision allows the production of an overall wiring diagram complete with details of the format of each termination enclosure, the design of each cable and the jointing, testing and commissioning on a route-by-route basis using the relevant nodal matrix (see Figure 11.5).

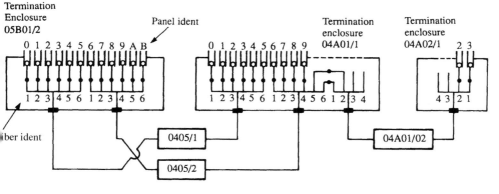

Figure 11.5 *Wiring schematic diagram*

However, the wiring diagram will define the desired number of optical interfaces (connectors, splices etc.) while the service requirements will define the bandwidths necessary within the installed cabling.

Optical fiber

Bandwidth is inversely proportional to light acceptance and interface losses. Put more simply, the ability to get light into an optical fiber improves as the core diameter and NA increase. So, in general, do the losses at the connectors (for a given connector design) and, to a lesser degree, the splice joints. However, bandwidth decreases with increased NA. As has already been stated it would be logical to install single mode technology in every environment if it were not for the cost of injecting light into the fiber. So, for the moment at least, compromises must be reached.

When viewed from the operational requirement the operating bandwidth must be the foundation upon which the design is built. Within the relatively short distances encountered in data communications (be they campus, backbone or horizontal links) the available bandwidth of optical fibers can be treated as behaving linearly with distance. Therefore the wiring diagram may be consulted together with the route description to determine the longest route. The operational requirement may then be consulted to determine the most demanding service with regard to bandwidth.

It cannot be assumed that just because an optical cabling system meets the design requirements of ISO 11801, or any other cabling standard, that the desired local area network will run over the desired distance. For example, 100BASE-FX, i.e. Fast Ethernet, is happy to transmit up to 2 kilometres over 62.5/125 multimode fiber; using ten gigabit Ethernet over the same fiber will result in a reliable transmission distance of just a few tens of metres. It is essential to consult the tables provided in Chapter 9 to ensure that power budget, fiber selection and LAN operation are concomitant with user expectations.

Having defined the bandwidth requirements of the optical fiber it is time to turn to the attenuation of the proposed configurations. Referring once more to the wiring diagram it is necessary to identify the link which will introduce the highest optical attenuation by virtue of either the connectivity or the length of the fibers within it. Calculations can then be undertaken to determine whether for the baseline performance fiber the proposed connectivity can be supported by the equipment to be used. If the attenuation is likely to be too great, then a reduction in the number of connectors must be considered by resorting to the use of splicing alone. This may reduce the desired degree of flexibility by the removal of patching facilities and an example of such a calculation is shown in Figure 11.6. A rather costly alternative is to introduce repeaters at convenient points along particularly difficult routes.

Specification definition

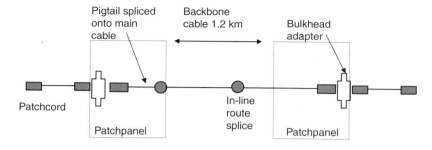

Figure 11.6 *Optical loss budget*

The above analysis must be made for each link to be offered on the highway. Equally importantly the calculation ought to be undertaken for all services predicted to be required. This ensures that the cabling installed will meet both initial and future requirements and that no surprises are to be found which would necessitate reinstallation at a later date.

As a result of these considerations it is not uncommon for an installed cabling infrastructure to include multiple optical fiber designs including both multimode and single mode variants.

Optical fiber cable

The route description should define the environment through which the cables pass. Any particular hazards should also be highlighted. For example, the petrochemical industry promotes the use of lead, or nylon, sheathed cables. Extreme conditions such as temperature, moisture or vibration should be highlighted within the operational requirement. This allows the installer to select and submit cable design proposals in relation to the conditions relevant to each route. This normally results in the suggestion of separate indoor and outdoor cable designs etc.

Demountable connectors and splices

The choice of demountable connector is relevant only when patching facilities are required. For multimode fiber designs the use of SC or ST connectors has been common whilst for single mode highways the FC/PC is dominant. The features and benefits of the various designs have been covered in earlier chapters; however, it is the performance of the connectors, rather than the design itself, which is relevant at the specification stage.

In the preceding section describing fiber choice it was the optical specification which defined the choice. Similarly the insertion loss (random mated) of demountable connectors is the key factor in determining whether a particular connector may be used or even whether a demountable joint can be accommodated within the optical power budget of the proposed transmission equipment.

In the same way the performance of joints such as mechanical or fusion splices must be defined.

Termination enclosures

Termination enclosures are as important to the eventual function of the cabling as are the passive optical components, and should be clearly defined at the specification agreement stage.

The style and position of the enclosures and, where relevant, cabinets containing the enclosures should be defined, including any aesthetic requirements. The design proposal should detail the method of cable management within the enclosures. This should include details of the glands, strain relief mechanisms and the fiber management systems. When the enclosures are to be fitted within cabinets, details of the cable storage should be submitted. The user should be satisfied that access to individual enclosures and individual fibers within them will be achievable without damage to other cables and fibers.

Optical specification

Having chosen the design of optical fiber, cables, demountable connectors and the methods of jointing them, a firm and fixed specification may be generated for the optical performance of those components. As will be seen in later chapters it is the installer's responsibility to install the individual components to these specifications and, by doing so, meet the overall optical loss specifications for the particular spans within the cabling infrastructure.

The optical specifications, with numbers supplied from ISO 11801, EN 50173 or TIA/EIA 568B, should include the following elements:

- *Fiber optic cable*   Fiber geometry (and tolerances)
    Fiber numerical aperture (and tolerances)
    Fiber attenuation (in cabled form)
    – 850 nm
    – 1300/1310 nm
    – 1550 nm (where relevant)

- *Demountable connectors*
- *Splices*

Fiber bandwidth (or dispersion)
 − 850 nm
 − 1300/1310 nm
 − 1550 nm (where relevant)
Random mated insertion loss
 − 850 nm
 − 1300/1310 nm
 − 1550 nm (where relevant)
Random mated return loss (where relevant)
 − 850 nm
 − 1300/1310 nm
 − 1550 nm (where relevant)
Random mated insertion loss (in final form)
 − 850 nm
 − 1300/1310 nm
 − 1550 nm (where relevant)

These specifications are vital in the production of an operating cabling system and as a result are the subject of continued observation during the installation phase.

Contractual aspects of the specification agreement

The foregoing sections describe the production of a basic operational requirement and the subsequent determination of a design proposal which includes the specification of the components to be incorporated within the cabling designed to meet that requirement. This represents the completion of the high-level technical specification and as such forms the first part of a specification agreement between the user and installer. As has already been said the establishment of the technical specification is the key as it sets down the agreed performance prerequisites for all the services to be operated on the highway. However, the bulk of the specification agreement concentrates upon the working practices and contractual issues which are necessary to ensure that the technical specification is complied with.

The installation programme

The specification sets out the contractual responsibilities of both user and installer in a manner aimed to ensure the smooth running of the installation.
 A typical installation will feature the following stages:

- Delivery of fixed cables (either to the installer's premises or to site). This occurs as the culmination of the following processes:
  − manufacture of optical fiber;

- shipment of optical fiber to the cable manufacturer;
- manufacture of cable;
- re-reeling of cable to specific lengths needed;
- shipment of cable to installer;
- shipment of cable from installer to site;
- Completion of any civil engineering works including provision of ducts.
- Completion of cable routeing tasks including the installation of traywork, trunking and other conduits.
- Laying of fixed cables (by the installer or the installer's subcontractors).
- Fitting of termination enclosures and cabinets.
- Attachment of fiber optic connectors to the fixed cables by the methods chosen in the technical specification.
- Testing of the complete cabling to prove compliance with the technical specification.
- Documentation of the completed cabling.

The above stages are all potential sources of contractual problems and Chapters 12–16 cover in detail the various methods of ensuring, in as far as possible, trouble-fee running of the installation contract once awarded. To this end the contractual aspects of the specification should be produced following the route outlined in those chapters. At the top level, however, are the general aspects of contractual practice which should be clearly stated and agreed at the outset between user and installer and detailed in the specification.

Contractual issues for inclusion within the specification

The following issues should be covered within the specification, which can also refer to EN 50174 (*Information technology, cabling installation*):

- scope of work;
- regulations and specifications;
- acceptance criteria;
- operational performance;
- quality plan;
- documentation;
- spares;
- repair and maintenance;
- test equipment;
- training;
- contract terms and conditions.

Scope of work. Whilst the operational requirement provides a technical definition of the task to be completed the scope of work defines the contractual boundaries of the project.

Misunderstanding or blatant disregard in relation to the responsibilities pertaining to the project are a common source of dispute (and delay). The scope of work includes clear definition of those responsibilities and defines the contractual interfaces between organizations involved in the various phases of the installation. The following details may feature in a scope of work section:

- task definition and boundaries;
- route information;
- survey responsibilities;
- bill of quantities;
- programme requirements and restrictions.

Regulations and specifications. All relevant general and 'site-specific' regulations should be clearly identified together with the health and safety regulations (both general and specific). All relevant materials and performance specifications should also be defined in accordance with, where necessary, the operational requirements already determined.

Acceptance criteria. The criteria to be used which will allow the user to accept the final cabling infrastructure must be defined. These criteria are not solely optical in nature. Indeed the majority of the tasks to be undertaken in a large installation project will relate to non-optical aspects. If staged payments are to be made to the organizations involved then it is necessary to define the criteria for each stage.

Nevertheless the optical acceptance criteria tend to feature strongly. The agreed optical performance specifications for attenuation etc. must be formally recorded and should not be subject to change without a contractual change being instituted. No work should be undertaken without full agreement to the viability of the acceptance criteria.

Operational performance. As has already been stated the cabling is designed to incorporate elements having predefined levels of optical performance which will be complied with by the use of the above acceptance criteria and by conforming to an agreed quality plan. However, the overall requirement is for operation of transmission equipment. Such equipment must comply with limits with regard to optical power budget and bandwidth set by the design of the cabling. These limits should be established, having given consideration to repair and reconfiguration of the cabling. Any relevant environment factors such as temperature should also be taken into account. This results in a power–bandwidth envelope within which any equipment may be operated.

Quality plan. Within the invitation to tender the user should highlight the need for the installation to proceed using components and techniques of assessed quality. To ensure compliance with this philosophy the user may request the installer to prepare a quality plan which must include:

- planning documentation;
- acceptance tests (type, quantity and programme);
- final highway tests (type, quantity and programme).

Within the quality plan the installer must highlight any limitations in relation to the test methods proposed either by the user or the installer.

Documentation. There are many standards to which a particular cabling installation may be documented. The specification agreement should define the contractual requirements for the documentation not merely supplied to the user following completion of the task but also to be provided and used during the programme of installation.

Spares. The specification should define the desired level of spare components and assemblies following consultation with the installer.

Repair and maintenance. Having considered the necessity for repair contracts and maintenance cover, the requirements should be clearly defined in terms of response time and contract terms.

Test equipment. When the user has a requirement for similar test equipment to that used on the installation the specification should define that need and the level of training to be supplied in order that the equipment may be useful in its proposed role.

Training. There are many types of training that are relevant to the use of optical fiber as a transmission medium. These include:

- operation of equipment;
- operation of installed cabling;
- fault analysis (equipment and cabling);
- user-based maintenance.

The specification should define the training needs and state clearly the responsibilities for their provision.

Finally the overall terms and conditions of the contract must be clearly stated. Agreement to these is obviously key to the specification being agreed between the user and the installer.

Summary

The production of a comprehensive specification is vital to ensure the smooth running of any optical fiber cabling installation.

By documenting as many issues as possible following the establishment of a comprehensive design, the agreement of the specification enables discussion at a non-technical level of the points of concern with the installation.

The greatest benefit is that the use of a specification maximizes the chances of a user actually getting the product for which the money has been paid.

12 Acceptance test methods

Introduction

The optical content within the specification defines the optical performance of the components and techniques used to produce the optical fiber highway and the networks it supports. Quality assurance begins with the setting of performance requirements for the incoming components or subassemblies to be used. Failure, on behalf of the installer, to fully test or otherwise certify these items can lead to contractual problems both during the installation and, in the worst case, at the end of the installation (when the network fails to operate). Such contractual problems can lead to financial penalties and damage to reputations (both corporate and personal) and they can influence the acceptance of the optical medium in future installations. As a result it is preferable to minimize their likelihood via the adoption of a quality plan. A comprehensive quality plan is a sensible requirement within any invitation to tender.

A typical installation includes the contractual steps shown in Figure 12.1. It will be seen that the key components to be included within the quality plan are the fixed cables and the various pigtailed, patch and jumper cable assemblies and the connectors applied to them. The termination enclosures should also be considered within the quality plan.

Fixed cables

By their very nature fixed cables are prone to contractual problems. The cable manufacturer procures the optical fiber and processes it into a cabled form. It is reeled onto a large drum during the manufacturing process and is then shipped to the installer or to the site of installation. If, however, the cable is held as a standard stock item and the installer requires only a short length, then the cable is re-reeled prior to shipment.

220 Fiber Optic Cabling

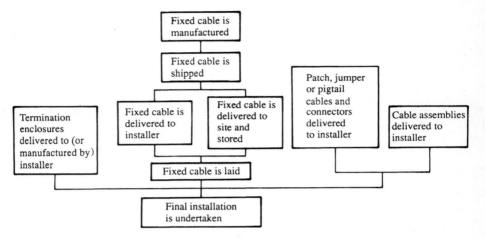

Figure 12.1 *Contractual phases with a typical installation*

Following delivery the cable may be stored at the site of installation before being laid and cut into the final installed lengths. These laid lengths are then joined and terminated to provide the final cabling networks.

Throughout this complex path a large number of organizations may become responsible for the fixed cables. These include the manufacturer of the optical fiber, the manufacturer of the fiber optic cable, the shippers, the cable-laying contractor, the installer of the final highway and, finally, the customer. With the product passing through so many contractual hands it is necessary for testing to be carried out at regular intervals to ensure that any damage or other deviation from specification may be highlighted, responsibility allocated and action taken.

Should the fixed cable not be to specification it is better for this to be determined at the earliest stage in the installation path since remanufacture or reinstallation will be expensive and may involve extended time scales. The expense of such remedial action must be borne by one of the parties involved. If a quality plan has not been produced, and complied with, it may be difficult to allocate blame, and as a result the price may have to be paid by an innocent party.

Fixed cable specification

Fixed cable (and any other type of fiber optic cable) should be purchased against a specification. The content of the specification divides into physical and optical parts. The physical part comprises the mechanical aspects of the cabling design, whereas the optical part details the design and desired performance of the optical fiber within that cable.

The mechanical aspects of the specification are unlikely to be affected during the installation programme. Obviously the cable might become damaged during the laying phase but this possibility will be minimized by the correct choice of cable sheath and by the adoption of the correct laying procedures.

The optical aspects of the specification are split between the basic physical parameters of the optical fiber, which cannot change during the installation, and the operational parameters which may be affected during the installation.

Acceptance testing of fixed cable

The fixed cable is normally delivered by the shipper to either the installer or direct to site. It is vital that the cable documentation is studied at this stage and accepted against the physical and optical specification to which it was purchased.

At this stage large cables will be on a drum and normally only one end of the cable is accessible. The cable drum should provide adequate means of transport and storage for the cable and normally features battens or similar to protect the outer layers of cable from damage. Additionally the cable ends should be protected from ingress of moisture and contaminants by the use of end caps (either heatshrink or taped designs).

During the fixed cable acceptance the physical aspects of the cable are checked off against the specification. The key issues are as follows:

- *Cable design, cable materials and markings.* Loose or tight construction, quantity of tubes (or fillers), quantity of fibers within tubes, tube and fiber identification, sheath materials, presence of correct type of moisture barrier, presence of correct type of strength members and presence of correct sheath markings and labelling.
- *Damage to cable sheath (outer layers on drum).* Sheath defects such as pock marks or cuts may be caused during production or shipment. It is essential to inspect the battens and drum walls for damage prior to testing. Although it is only possible to inspect the outer layers of the cable on the drum, the cable laying contractor should be made responsible for inspection of the cable as it is removed from the drum.

With regard to the optical specification it is important to obtain (by detailing the requirement within the purchase order) the following:

- Certificates of conformance for fiber geometry and numerical aperture (at the desired operating wavelengths).
- Certificates of conformance for fiber bandwidth (at the desired operating wavelengths).
- Certificates of conformance for refractive index of the optical core (at the desired operating wavelengths).

None of the above four parameters is easily measured in the field and documented statements of compliance with specification are necessary and should be kept for future reference. It will be noticed that the parameters relate to the optical properties of the optical fiber as it was manufactured and the cable manufacturer should, providing that levels of quality assurance are adequate, be able to supply records of the measurements made by the optical fiber manufacturer should that become necessary.

For large and expensive cables it may be appropriate to measure and test each fiber before the cable is installed, so if the cable subsequently fails after installation, then it will be apparent that the problem occurred during the installation. If this is not done then there will always be an argument between the installer and cable manufacturer as to whether the cable was delivered in a working order.

The final section of cable acceptance testing relates to the optical aspects of the cable construction. It is desirable to know the length of each fiber element within the cable (and to know that all of the elements are of the same length). Furthermore the attenuation of the optical fiber elements within the cable must be deemed to be within specification and should not show localized losses consistent with applied stresses (either during production or reeling). For this the installer must use an optical time domain reflectometer (OTDR).

The optical time domain reflectometer (OTDR)

An optical time domain reflectometer is possibly the most useful analytical tool available to the installer. It can be used to perform inspection and testing of fiber optic cables of all types and lengths. The hard copy results produced can be included in contract documentation and represent performance baselines against which subsequent measurements can be compared. But as a word of caution, many installers do not know how to properly use an OTDR and even if the results are meaningful they still require expert interpretation.

An OTDR may be used to test completed networks and provides a remarkably accurate assessment of the individual attenuation levels produced at the various joints and demountable connections throughout the cabling. For short length optical projects, however, i.e. less than 2 km of multimode or 1 km of single mode, the OTDR is really an overkill for simple acceptance testing. A power meter and light source is best for acceptance testing in projects of the aforementioned size, but the OTDR remains the ultimate fault-finding tool.

As a means of assessing long fiber optic cables the OTDR is invaluable since it can detect and locate specific localized attenuation events consistent with applied stress in addition to measuring the length and overall attenuation of the individual optical fiber elements.

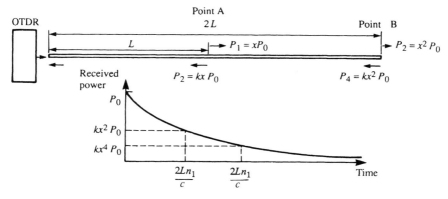

Figure 12.2 OTDR theory

The OTDR operates by launching a short pulse of laser light into the optical fiber to be measured. This light is scattered (by Rayleigh scattering as discussed in Chapter 2) at all points along the fiber and a small fraction is scattered back towards the OTDR. The backscattered light is captured by the OTDR and analysed to produce an attenuation profile of the optical fiber along its length.

With reference to Figure 12.2 two points on the optical fiber element are considered. The power launched into the optical core by the OTDR P_0 mW will be attenuated by the optical fiber until it reaches point A. The forward power at this point is $P_1 = xP_0$ (where $x < 1$). The scattering fraction will be small (and dependent upon wavelength of transmitted light) and is normally constant within a given batch of optical fiber. The light scattered back towards the OTDR can be written as $kP_1 = kxP_0$ at point A. This light is also attenuated as it returns to the OTDR and the light reaching the OTDR is $kxP_1 = kx^2P_0$.

Looking at point B the same argument applies but the values of received light are different. At point B, $P_2 = x^2P_0$ the scattered power is kx^2P_0 and the received power at the OTDR can be shown to be kx^4P_0.

To summarize, the scattered light powers received back at the OTDR area:

Point A: distance = L power received = kx^2P_0 time elapsed = $2xn_1/c$
Point B: distance = $2L$ power received = kx^4P_0 time elapsed = $4xn_1/c$

The elapsed time between transmission of the laser pulse and the reception of the scattered light can be calculated by the distance travelled divided by the speed of the light in the optical core.

By sampling the received power at predefined time intervals the OTDR is able to measure the reduction in power with increasing distance along the optical core. If the OTDR is given the refractive index of the

224 Fiber Optic Cabling

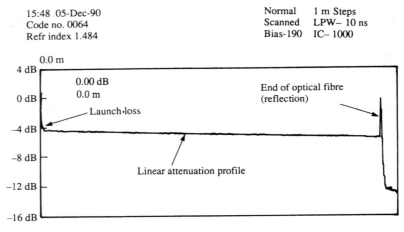

Figure 12.3 *Typical OTDR trace*

core material (required on the certificate of conformance from the cable manufacture) then the time sampling can be effectively converted into distance and the results observed will represent loss as a function of distance along the fiber.

Figure 12.3 shows a typical trace produced from an OTDR of an unterminated optical fiber within a cable. The launch loss produced by non-linear effects within the optical cladding near the OTDR dissipates and the trace becomes regular in form.

Using the OTDR the difference in received power between points A and B is shown above to be twice this figure; however, the OTDR converts all measurements to represent a single-way path (both in terms of attenuation and distance) and therefore the OTDR can be used to make representative measurements of the losses encountered in an optical fiber.

With reference to Figure 12.3 it can be seen that the end of the fiber is characterized by a large reflection peak. This peak is caused by Fresnel reflection as the light launched by the OTDR passes from silica to air. Using this peak the length of the individual optical fibers within a cable may be measured. Similarly by correct placement of the cursors a measurement of the attenuation of the element in dB/km may be made; however, care should be taken to avoid incorporating the launch loss within the calculation.

Measurements using the OTDR

Optical time domain reflectometers are designed for specific single or multiple wavelength operation and it is important to choose the appropriate equipment for the measurement to be made. To check the performance of an optical fiber against a specification at 850 nm an 850 nm OTDR must be used and the same is true in the second and third windows.

Also the length of the cables must be taken into account when choosing the correct OTDR for use. The original application of OTDR equipment was the location of faults in telecommunication system cabling. This meant that much of the earlier multimode equipment was designed to 'see' as far as possible from one end which was accomplished at the expense of resolution and dead zone (a measure of the length of the region immediately following a reflective event within which measurements cannot be made). The early first window equipment had ranges of 10 km, resolutions of 10 m and dead zones of 10 m. More recent equipment, specifically designed for the local area network, readily achieves ranges of 5 km but with significantly improved resolutions (0.5 m) and dead zones (2 m). This type of equipment is ideal for all normal cabling installations and allows measurements on cables as short as 20 m in length.

As progress was made into the second window new, more sensitive equipment had to be developed since the Rayleigh scattering falls away rapidly with wavelength. Furthermore the move towards single mode with its markedly reduced launch powers compounded the problems. As a result second and third window equipment (whether multimode or single mode) tends to be more expensive than first window, multimode machines. Also the telecommunications dominance of these windows and fiber designs makes the OTDR equipment not as applicable to local area network cabling.

In most countries it is possible to get an OTDR calibrated in accordance with national or international standards. Such calibration is important to ensure repeatability of measurements made and is frequently, if not always, a stated requirement of a specification and the accompanying quality plan, and the user may seek documentary evidence of the calibration status of the equipment used.

Large external cables delivered on drums may be tested with an OTDR before any further processing takes place. The cable end cap should be removed and the cable stripped down to expose between 0.5 and 1.0 m of usable fiber lengths. Care should be taken to avoid damaging any of the elements within the cable construction. A temporary termination may then be applied to each of the elements in turn, connected to the OTDR and the measurement made.

The aspects of the specification which can be checked using the OTDR are shown in Figure 12.4. Having provided the OTDR with the relevant refractive index, the length of cable should be checked to be in line with the installer's requirements and all fibers within one cable construction should be checked to ensure that all have the same length (within

226 Fiber Optic Cabling

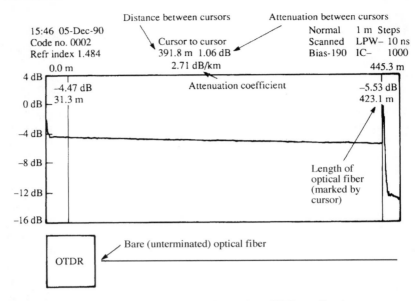

Figure 12.4 Cable acceptance testing using OTDR methods

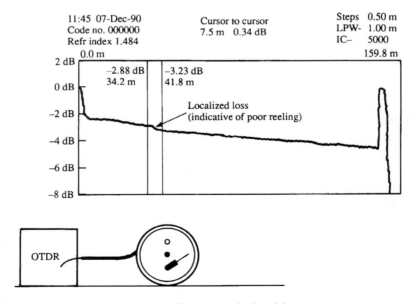

Figure 12.5 Localized attenuation on reeled cable

Acceptance test methods 227

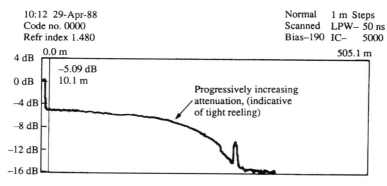

Figure 12.6 *Major attenuation problems within reeled cable*

measurement accuracy). Any other result would suggest a break in one or more elements. Attenuation should be in line with the purchase specification and the trace should be uniform. Any localized deviations (see Figure 12.5) must be investigated since they could be indicative of stress applied to the individual optical fibers during the cable manufacturing process and, as a result, might suggest early failure of the elements once installed. Alternatively, such localized losses might have resided within the optical fiber before cabling and might actually lie within the specification of the fiber itself. In this case documentary evidence may be sought from the cable manufacturer in order to prove that the loss events have not been exacerbated by the cabling process. Other, more systematic, loss features (see Figure 12.6) should be investigated in the same way as detailed above.

Once measured the trace for each fiber element must be stored and transferred to hard copy (or electronically) for inclusion as part of the final documentation.

Acceptance testing of laid cable

Once the cable is laid, the need to test the fixed cable is paramount. The reasons for this are straightforward and can be summarized as follows:

- As the laying process frequently represents the first time the cable has been unreeled since manufacture and delivery it is vital to establish any differences in performance between the reeled and unreeled states.
- Once laid, the fixed cable assumes its final configuration as part of a complete infrastructure. As such its performance is a component part of that infrastructure and it merely remains for it to be jointed or otherwise terminated to become a fully functional optical fiber span. It is therefore important to be fully aware of any non-compliant aspects such as localized losses before proceeding to the final installation phase.

- The laying of fixed cable offers maximum opportunity of damage, particularly via kinking, twisting and bending. Also the possibility of third party damage is increased during or after the cable is laid. Obviously the correct choice of fixed cable will minimize the chance of damage during the laying of the cable provided that sensible procedures are adopted. Nevertheless human error is always a possibility and testing is vital for certification purposes.

The OTDR is once again invaluable in identifying breaks in individual fiber elements and localized losses due to stress applied to the cable during the laying phase. Long lengths, such as seen in telecommunications projects, would be carried out on unterminated fiber immediately before the fixed cable is glanded into the termination enclosures. The cable sheath and moisture barriers should be removed over a length of approximately 1.0 m from each end of the fixed cable and a temporary termination applied to each element in turn.

The OTDR results should show that all of the optical fibers within the fixed cable are of the same length from both ends since any other result would imply a break. The attenuation per kilometre (if measurable: short lengths of cable below 100 m can give misleading measurements) should be in broad agreement with the figures taken whilst the fixed cable was still on the reel or drum. But, most importantly, there should be no

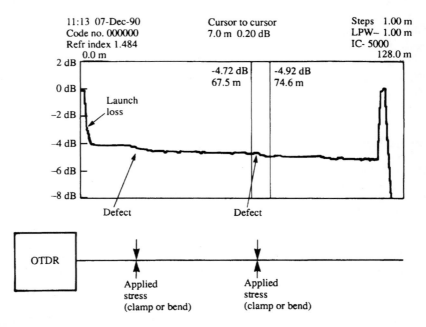

Figure 12.7 *Localized attenuation in laid cable*

localized losses which were not present at the time of delivery. Presence of such losses (see Figure 12.7) would suggest some type of applied stress leading to light being lost at the CCI. These must be investigated and removed, otherwise the chances of future catastrophic failure of the affected optical fibers may be significantly increased.

Air-blown fiber testing

Air-blown fiber consists of empty plastic tubes, generally 5 to 8 millimetres in diameter, into which bundles of fibers or individual optical fibers are blown in using compressed air, or other gases, at a later date. Once the optical fibers have been installed they are terminated and tested in exactly the same manner as any other kind of optical fiber. The blown-fiber ducts, however, need a particular kind of test after installation to ensure they have not been damaged, which would of course preclude any fiber blowing in the future.

The blown-fiber duct test is in two parts, but it is conducted by one machine or device. A device injects a steel ballbearing into one end of the duct under pressure from compressed air. If the ballbearing makes it to the special valve at the far end then the installer knows that the duct has not been crushed or kinked. The specially designed valve at the far end is sealed by the arrival of the ballbearing and the pressure in the duct will start to rise if the source of compressed air is still attached. The duct is allowed to rise to its normal maximum working pressure, typically ten times atmospheric pressure, and if the duct maintains that pressure over a few minutes then one knows that the duct has not been punctured or ripped anywhere. The duct is then sealed at both ends to prevent the ingress of water, dust and insects, and it should be possible to return to that duct at any time in the future to blow fiber. If a long time has elapsed, however, then it would be wise to pressure test the duct again to ensure that no physical damage has happened to the ducts in the intervening years. If the ducts are damaged then the optical fiber will not blow into the duct or it may enter the duct only to stop at the point of damage, and the fiber consumed would normally be lost.

Very short runs, as may be encountered in fiber-to-the-desk, may not warrant the added cost of pressure tests as the chances of problems are very slim.

Cable assembly acceptance testing

Fiber optic cable assemblies can be regarded to be either pigtailed, jumper or patch cable assemblies. These tend to be manufactured in factory conditions and are therefore purchased as finished items against a given specification. It is a regrettable fact that many installers do not give due consideration to these components and subassemblies. Only when problems occur do the installers begin to appreciate the importance of quality assurance as it applies to the terminated cable. It should be emphasized that the terminations applied to the fixed cables, jumper and patch determine the performance of the network and without good terminations it may be impossible to inject light into the cable or impossible to patch between termination enclosures, thereby rendering the entire network inoperable.

Chapters 4 and 5 discussed the issues relating to the jointing, by various means, of optical fibers. With particular reference to demountable connectors, Chapter 5 introduced the concept of insertion loss and return loss. The terminations applied to fiber optic cables must each meet an agreed specification, of which insertion and return loss represent the optical aspects. There is also a physical specification to be complied with relating to the mechanical (dimensions, strength etc.) and the surface finish (fiber/connector end-face) aspects of the cable assembly.

Optical performance

The quality of a particular termination can be assessed only by a method that measures its specific performance against a similar termination. The factory producing terminated assemblies such as pigtailed, patch or jumper cable assemblies can only truly assess their abilities by taking a measurement of the attenuation introduced within a demountable joint using each termination in turn. This is defined as the insertion loss for the joint.

The return loss measurement techniques vary but they refer to specific terminations or joints rather than assemblies.

Insertion loss

Insertion loss is normally measured in the factory using stabilized light sources, operating at the desired wavelength and incorporating the correct type of optical source, together with optical power meters, incorporating the correct design of detector (matched to the wavelength of the power source). Figure 12.8 shows the typical arrangement.

The insertion loss of the terminated connector must be measured against an identical connector within an adaptor produced by the same manufacturer. This, by definition, should give some degree of consistency since manufacturers' specifications are virtually always written around a consistent component set.

For this reason it is necessary to introduce the concept of launch leads which are made from the correct fiber type, i.e. to the same specification as that used within the cable to be tested. It is preferable for the launch

(a) Setting reference measurement

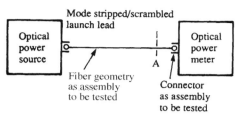

Optical power source shall
- operate at correct wavelength
- be relevant to operating system (multimode LED, laser; single-mode laser)

Optical power meter shall
- operate at correct wavelength

(b) Making the measurement

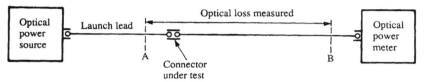

Figure 12.8 *Insertion loss measurement of terminated cable*

leads not to be manufactured from the same batch of optical fiber since it does not always introduce the desired amount of randomness which may be encountered in the installation environment. Nevertheless the fiber should be to the same specification.

The launch lead should be terminated at one end with a connector suitable for connection to the power source and at the other with a connector of the same design as the termination to be tested.

The launch lead must be cladding-mode stripped and core-mode scrambled. The first can be achieved either by the use of an applied cladding mode strip or by the use of a long length of cable. An applied cladding-mode strip takes the form of a material of high refractive index directly bonded to the surface of the cladding along a prepared length (normally between 100 and 150 mm) of the cable. This removes the optical power launched by the source into optical cladding of the launch lead.

Mode scrambling involves the production of a launch condition emitted from the launch lead in which the power is distributed across all the possible modes within the optical fiber. This most closely represents the ideal transmission condition within the installed cabling which includes joints over a considerable length. Scrambling is achieved by the introduction of a tight mandrel wrap on the cable or by using a long length of cable, e.g. 5 turns of 62-/125 tight-buffered fiber around a 20 mm mandrel.

The power meter is fitted with an alignment adaptor which is of the same generic type as the connector to be tested. These adaptors are simply responsible for targeting the light within the optical core onto a consistent area of the detector surface. The detector is normally very much larger than the core area and therefore micron level alignment is not important.

Measurement method

The launch lead is connected between the power source and power meter. The detector within the power meter 'sees' all the optical power within the core (less a small but consistent amount of Fresnel reflection) and can be represented diagrammatically as being the power at point A, just behind the connector.

The power meter is 'set to zero' making this measurement a reference value. The launch lead is then disconnected from the power meter and the test termination connected to it using the correct adaptor.

If the termination at the opposite end of the test lead differs from that under test the power meter adaptor should be changed to suit. This is true even if the test lead is a pigtailed cable assembly since bare fiber adaptors are available (however, it is easier to terminate both ends of a cable, test as a patch cable assembly, and then cut it in half).

The free end of the test lead should then be connected into the power meter and the measurement noted. As the detector 'sees' all of the light in the optical core (with the exception of the same Fresnel reflection mentioned above) then the measurement made is of the increase in attenuation produced by the insertion of the section AB. Providing that the cable length of the test lead does not contribute significantly to the loss, then the measurement is directly related to the insertion loss of the demountable connector joint.

When the connectors used to be able to exhibit rotational variations (e.g. SMA 905 and 906) then this method was sometimes extended by making multiple measurements and the mating connectors were rotated through 90, 180, 270 and 360° against each other.

The validity of the insertion loss measurement must be fully understood. It does not represent the only value which could be achieved from a demountable joint containing that test termination since, as was discussed in Chapter 5, the loss depends not only upon the physical alignment of the fibers but also upon the basic parametric tolerances of the fibers within each connector at the joint. Therefore a different launch lead could and would produce a different result.

So what is the manufacturer looking for in an insertion loss measurement? Referring back to Chapter 5 it was stated that the relevant figure for a demountable connector pair was the random mated insertion loss and the maximum value must take into account the basic parametric tolerance losses and the quality of alignment of the particular connectors

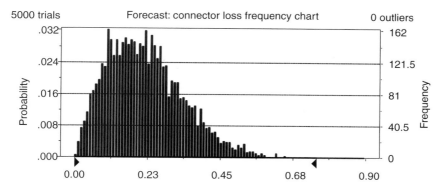

Figure 12.9 *Insertion loss histogram*

used. There is a maximum random mated insertion loss for every connector design on a specific fiber design. Statistically most demountable connector joints will exhibit insertion losses below the theoretical maximum when measured in the above manner against a randomly selected launch lead. Against a different launch lead the demountable connectors will continue to exhibit statistically similar results but individual results will differ. Therefore the insertion loss measurement is not an absolute value but is merely confirmation that a given test termination falls within the statistical distribution bounded by the maximum random mated insertion loss figure.

To summarize: provided that the insertion loss measurement lies within the agreed random mated insertion loss defined within the specification agreement, then it is acceptable; however, its performance is assumed to be at the limit, i.e. the maximum random mated insertion loss. Figure 12.9 shows a typical insertion loss histogram for optical connectors.

As an alternative to insertion loss testing of individual components it is possible to test complete jumper or patch cable assemblies to assess the similarity of their overall performance rather than the performance of the individual terminations. Whilst this is rarely undertaken outside the telecommunications industry it is included for completeness.

Substitution loss measurement

Using this measurement technique jumper or patch cable assemblies are subjected to comparative measurements. The test procedure is shown in Figure 12.10.

For these measurements a launch lead and tail lead are produced using the same design of optical fiber as that of the assemblies under test. As for the insertion loss method the launch and tail leads must be sufficiently

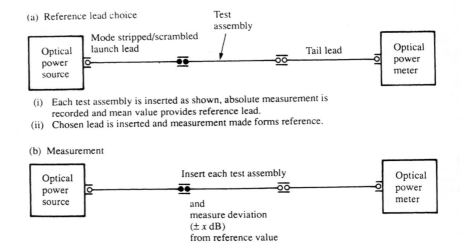

Figure 12.10 *Substitution loss measurement*

long to remove all optical power injected by the source into the optical cladding and fully fill the available modal distribution within the optical core.

The first test lead is connected between the launch lead and tail lead and the measurement on the power meter taken as a reference. Each test lead is inserted, in turn, and the deviations from the reference value recorded. Once all the test leads have been measured the mean deviation is calculated. The test lead most closely matching that deviation is then chosen as the reference lead.

The reference lead is then reinserted in between the launch lead and the tail lead and the measurement taken as the true reference value. The remainder of the test leads are once again measured in turn and the deviations recorded. These deviations are compared with an agreed specification and the acceptable cable assemblies are then considered to have an equivalent attenuation.

Return loss

The return loss of a particular mated connector pair is of special concern to the installers of cabling structures in which lasers are to be used. As has already been said the performance and lifetime of many laser devices is dramatically affected by light reflected back from within the optical fiber. The reflections are caused principally by the Fresnel effect and the development of physical contact connectors such as the FC/PC was undertaken to reduce the level of reflected power.

For this reason most single mode connector designs have a specified return loss which is tested either on a 100% or sample basis. The main markets for such connectors are the various telecommunications and CATV organizations which tend to have individual corporate standards for the measurement although generic standards are now available.

Visual acceptance of cable assemblies

The methods of testing the optical performance of terminated cable assemblies discussed above are not the only techniques but they are those commonly used in volume production facilities. As a result they are the methods with which anyone purchasing assemblies should be aware.

However, optical performance is not the only criterion for acceptance of cable assembly products. Initial optical performance is no guarantee of operational lifetime and it is therefore necessary to define a visual inspection standard.

The visual inspection obviously includes the physical parameters such as length, type of cable, relevant markings and labels together with the type of connectors applied, but it goes much further than that. It is generally accepted that the majority of all faults in installed cabling result from faulty or damaged connectors. This is not surprising since the demountable connection is accessible to both authorized and unauthorized fingers and contamination can result. It is important then to ensure that the initial condition of the demountable connector end-face is such that premature failure will not occur due to built-in stresses and other forms of damage linked to poor standards of manufacture.

Pistoning effects

The physical quality of a termination can be measured in terms of the fiber end-face itself (discussed below) and the stability of the fiber within the connector.

The stability of the fiber is determined by the bond created between the fiber reference surface (i.e. the surface of the cladding) and the connector ferrule. One of the greatest disadvantages of the crimp–cleave or dry-fit connectors is that the bond is not created by an adhesive and stability is poor. This can result in the fiber end growing out beyond the front-face of the ferrule as the various cable components, such as the plastic sheath material, shrink back over a period of time. This effect was seen on a frequent basis when plastic clad silica fiber designs were used in the early years of optical fiber in the non-telecommunications area. Fiber grow-out became a well-known phenomenon with protruding lengths of many millimetres not being uncommon. Damage to mating connectors and equipment was widespread and retermination of the cables was necessary but was no solution since the effect simply recurred.

Pull-in is less common but can and does occur when the fiber appears to retract inside the ferrule. This is due to expansion of the cabling components.

Both pull-in and grow-out are commonly termed pistoning effects.

The use of adhesive based or epoxy–polish connectors significantly reduces the possibility of pistoning because the cladding surface is firmly bonded into the ferrule, normally over an extended distance. Professional termination facilities achieve very high yields by choosing the correct type of adhesive for the particular connector and fiber design and therefore bond failure is rare. However, poor processing can lead to bond failure and insufficient cleaving of the fiber surfaces or incomplete curing of the epoxy resins used can lead to major problems.

Large geometry fibers (100/140 microns and above) tend to exhibit greater shear stresses at the cladding interface and pistoning effects are more common on these fibers. It is possible to accelerate such a failure by subjecting the completed termination to thermal cycling (−10 to +70°C). This has a cost impact, but is a worthwhile expense, easily justified for the larger-core fiber designs when the alternative of damaged connections or equipment is considered. For fibers with 125 micron cladding diameters the likelihood of pistoning effects is considerably reduced, providing correct processing methods are used, but to provide a certified product release thermal cycling may be deemed desirable.

Surface finish

Surprisingly the surface finish of terminated optical fiber end-faces does not always have a drastic effect upon the optical performance of the subsequent joint. As a result there are few agreed inspection standards. This text adopts the inspection standards which have become *de facto* in nature throughout the termination industry.

All professional connector designs involve the polishing of the fiber end-face as shown in Figure 12.11. The methods of polishing vary according to manufacturers' recommended instructions; however, it should be pointed out that the connector manufacturer may indicate techniques that suggest faster throughput (to aid the acceptability of the product in the market place) which are not in the long-term interest of the terminated product.

For the purposes of visual inspection the end-face of the ferrule can be divided into the following regions:

- ferrule;
- face core;

Acceptance test methods 237

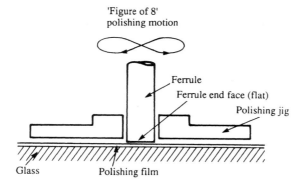

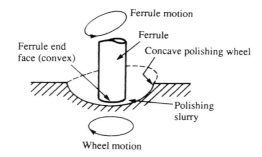

Figure 12.11 *Polishing of terminated optical fiber*

- inner cladding;
- outer cladding;
- adhesive bond.

Figure 12.12 shows these regions of which the core, bond and ferrule are self-explanatory and fixed. The cladding is divided into two separate regions, inner and outer. The position of the dividing line between the two regions is normally accepted to be midway between the core and cladding surfaces. The visual inspection under magnifications of 200 times or greater is based upon this definition of the regions of the connector end-face, and the acceptability of defects depends upon the region in which they are located.

The core and cladding regions may exhibit the following types of defect:

- *scratches*: surface marks consistent with external material being trapped on the fiber end-face during polishing or use;
- *cracks*: either apparent at the surface or hidden beneath the surface the

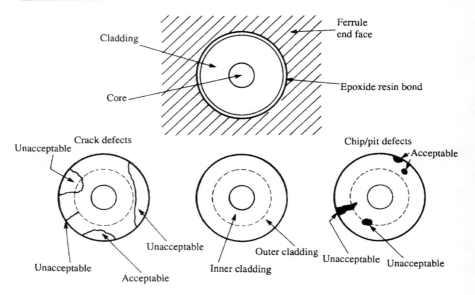

Figure 12.12 *Termination regions and visual inspection criteria*

cracks may be linear or closed and are caused by internal stresses within the fiber or by poor processing;
- *chips*: areas where a crack has resulted in the fiber surface breaking away;
- *pits*: localized chipping which may be extensions of scratches or consistent with the existence of internal stresses.

The most important inspection criterion is universal – there cannot be any cracks, chips or pits in the core region and any scratches seen in the core region can only be accepted if they are consistent with the means of polishing.

The inner cladding region should be similarly free from defects.

The outer cladding region can exhibit certain types of defects under certain conditions. First, it should be stated that some level of chipping and cracking around the edges of the fiber is inevitable and that the desire for no defects at all would be hopelessly impractical. The key issue is the absence of defects that will contribute to optical and mechanical failure of the termination during use. To this end the guidelines below have been developed.

In the outer core region closed cracks and chips are acceptable provided that they do not extend for more than 25% of the cladding circumference. Linear cracks are not acceptable, particularly if they may move toward the core in the future.

As was detailed above, the position of the division between the inner and outer core regions is normally taken to be the midway point. This general guideline is based upon a great deal of experience of crack failure mechanisms but it is inevitable that some customers may wish to move the division out towards the cladding surface, believing that improvements in overall quality will be gained. The additional cost incurred by the associated reduction in production yields is not, in general, justified.

Inspection conditions

Surface features such as scratches, chips, pits and certain types of crack may be clearly seen under magnification of 200 times using front illumination. However, buried cracks, which are potentially damaging should they spread towards the surface, cannot always be identified in this way and rear illumination is necessary. This is normally achieved by injecting visible light from the remote end of the cable assembly.

As a further level of quality assurance, thermal cycling or thermal shock may be used to accelerate any hidden cracks or stimulate any built-in stresses.

Direct termination during installation and its effect upon quality assurance

The inspection conditions outlined above are most easily provided in a purpose-built facility. For example, the ability to undertake thermal testing (cycling or shock) is severely restricted. Indeed, to inject visible light from the remote end of the cable, in order to detect hidden cracks etc., can prove rather difficult in an installation situation and if the installed link is long the selective attenuation of visible wavelengths may render any attempt futile.

Nowadays the vast majority of multimode connectors in premises cabling environments are field-installed, and the achieved quality is acceptable. Single mode connectors are still usually factory terminated onto a pigtail and the pigtail is fusion-spliced onto the main cable on-site.

Termination enclosures

The acceptance of termination enclosures is a combination of acceptance of the physical aspects of the enclosure itself together with the confirmation that the optical aspects of the enclosure will meet the requirements of the specification agreement.

The obvious physical aspects are the dimensional parameters, the material specifications and the presence of the correct glands, fiber management systems, safety labels and other markings.

If the termination enclosure is in a patch panel format, then the adaptors should be checked for fit and for compatibility with the connectors to be used on the terminated cable assemblies to be connected through the adaptors. In addition all adaptors must be provided with dust caps.

Pre-installed cabling

As time passes the amount of fiber optic cabling installed increases and installers will be called upon to extend or modify existing cabling infrastructures. It is important that installers should not take on any contractual responsibility for pre-installed cabling without first undertaking an inspection.

One major concern is that of optical performance and the compatibility between the existing components (such as fixed cable) and those to be installed. The installer must be satisfied that all demountable or permanent joints can be performed in line with the attenuation figures outlined in the specification.

Equally important is the matter of existing documentation. It is natural to accept existing documentation as being correct but if there is a doubt, no matter how small, the existing cabling should be surveyed prior to acceptance of the contract.

Short-range systems and test philosophies

The use of optical fiber has penetrated all forms of communication. In the early years long-range telecommunications dominated and more recently this has been followed by the growth of data communications in the inter- and intra-building markets. The dimensions of these applications are greater than the lengths required by the optical fibers to operate in a linear fashion. That is to say that the light can be considered to be travelling only inside the core of the optical fiber and exhibits a stable modal distribution. These two conditions are those met by the launch leads and tail leads used in the measurement of insertion or substitution loss earlier in this chapter.

The length taken for a particular optical fiber to reach these conditions is termed the equilibrium length of the fiber. This equilibrium length varies according to the type of optical source used, the fiber geometry and the physical configuration of the cable. For instance, the presence of a mode scrambler and cladding mode stripper will significantly shorten the equilibrium length. The presence of connectors and joints modify the modal distribution within the optical core and tend to scramble the modal populations.

If the system installed is of the order of the typical equilibrium length for the optical fiber used, then great care must be used in the assessment of component acceptance criteria, which are then compared with final system tests. This is because the performance of the components used is influenced by the style and type of optical components used within the transmission equipment.

The light injected from an optical source into an optical fiber experiences two forms of non-linear behaviour close to the point of injection. First, light is injected directly into the optical cladding due to misalignment of the fiber with the source. This light will be lost over a relatively short distance. This results in short lengths of optical fiber appearing to be more lossy, per metre, than longer cables in which the attenuation processes have settled down. Second, the modal distribution of light within the optical fiber can create additional loss characteristics which are dependent upon the type of device used. A small, low NA source such as a laser will launch light into the fiber with relatively few populated modes. As the light travels down the fiber the modal population will build up until the predicted modal content is achieved. Over this distance the loss of light from the core to the cladding is higher than in a long length as excess modes are generated and then stripped away.

Normally this is unimportant since it is normal practice to measure the power output from an optical source into an optical fiber which is cladding-mode stripped and mode scrambled. This is the basis for the determination of coupled power into an optical fiber as discussed in Chapter 9. All other components are measured using similarly conditioned test leads and therefore all results are compatible.

On short systems the testing philosophy may have to be modified to allow for the non-linear losses. The impact of systems with lengths shorter than the equilibrium length in the particular optical source and fiber used lies in the assumptions which must be made at the design stage. If the installed system is unlikely to produce a mode-stripped and scrambled-modal distribution, then any measurement using such conditions may be unrealistic. It may be more relevant to measure optical output power using a very short unconditioned test lead. It will be equally valid to measure the various cable assemblies using an optical source of the type used in the final system, again without the need for conditioned launch and tail leads. The immediate result is to improve the measured power output but to increase the measured attenuation of the intervening cabling.

To summarize: it may be appropriate to test a short-range system as a system rather than a set of independent components.

Further problems may arise where passive components are inserted into this system which have optical performance parameters that vary with modal distribution. This is a relatively unexplored topic but is relevant to the forthcoming short-range systems which depend upon branching devices (or couplers) for their operation.

13 Installation practice

Introduction

It has been a sad fact that some fiber optic cabling installations have been seen to have problems which have reflected badly upon the technology. In many cases the problems have been contractual in nature rather than technical, even though the symptoms may be technical (from the point of view that the cabling does not function). It has already been stated that the existence of a specification is vital to define the task to be accomplished. That being said the handling of the various contractual interfaces within the installation must be defined to ensure free running of the contract. Before considering these issues a number of terms must be defined.

For the purposes of this chapter the term 'installer' relates to the specialist fiber optic cabling installer (even if the cable is physically laid by a third party).

The term 'customer' relates to the organization having placed a contract upon the installer to provide services defined within the specification. The installation of an optical fiber data highway may be just one part of a wider communications system being established and a prime contractor may be responsible for the overall task. In this case the term 'customer' relates to that prime contractor.

The term 'third party' relates to any organization other than the customer and installer contracted either by the customer or installer to perform certain tasks within the requirements of the overall installation contract.

Successful completion of the installation depends upon the correct management of the contractual interfaces between customer, installer and all third parties. From the installer's point of view the possible interfaces are where the customer or a third party may:

- purchase some or all of the fiber optic components;
- provide pre-installed or pre-purchased fiber optic components;
- provide documentation relating to the site (which may include a manufacturer's warranty);
- provide documentation relating to existing fiber optic cabling;
- undertake civil engineering works;
- undertake cable laying;
- undertake installation of cabinets and/or termination enclosures;
- provide transmission equipment.

The adage 'trust nobody' holds true in any contractual interface and fiber optic cabling is no exception. This puts pressure on the installer and customer to accept nothing without valid acceptance test results or other documentary evidence.

Transmission equipment and the overall contract requirement

The customer may require the installation of a turnkey system including the fiber optic transmission equipment needed to provide the communication services initially desired upon the data highway. However, it is unlikely that the average installer will wish to take responsibility for the electro-optics at either end of the installed cabling. It is worthwhile explaining this reluctance from the technical viewpoint.

Throughout this book it has been stated that the two primary causes of the malfunction or non-function of a fiber optic transmission link lie within either the transmission equipment or the cabling. The cabling can only fail because of insufficient bandwidth or excessive attenuation. In practice the bandwidth of the installed cabling is unlikely to change and failure may be traced to attenuation at demountable connectors, joints or within the cables themselves. The installer therefore is able to quantify the performance of the installed cabling at the time of installation and at any time thereafter. Transmission equipment, however, can fail for a variety of reasons ranging from blown fuses to corrupted electro-optic signals being generated, and in general these causes are well beyond the capability of the installer to detect and rectify without specialist training. It is not unsurprising to find that installers are reluctant to take responsibility for components outside their normal sphere of operation. If the customer wishes to continue with the turnkey approach then another route must be sought.

The candidates most able to service the turnkey solution are known as system integrators or sometimes value added resellers (VARs), who sub-contract the various tasks involved whilst providing a project management role and warranting the entire package.

If the customer has a specific application for the data highway and does not intend to upgrade, expand and evolve the services offered, then the systems integrator may be very successful in providing the correct solution. However, if the cabling is intended to support a range of equipment and services over its lifetime, then it may not be in the best commercial interest of the customer to use a system integration approach, since the customer may desire the highway to be warranted and maintained separately from the transmission equipment. For this reason an installer may be contracted to provide a fiber optic cabling infrastructure as a stand-alone item. In this case the equipment supplier will be charged with the task of providing a transmission link across an installed cabling network for which a bandwidth and attenuation have been defined within a specification agreement.

This chapter assumes that the installer does not provide the transmission equipment.

The role of the installer

Armed with a specification, which defines the operational requirement and the optical performance limits which must be met, and an effective quality plan, in terms of the correct form of acceptance testing for the individual components to be used, the installation of fiber optic cabling can be separated into the following tasks:

- civil engineering works;
- cable laying;
- erection of cabinets and termination enclosures;
- jointing and testing of laid cabling components and accessories.

The practices involved are covered in this chapter. Obviously once installed the highway must undergo final acceptance testing and then be fully documented. These tasks are covered in subsequent chapters.

In virtually every case the civil engineering works and cable laying practices are the same as should be undertaken for good-quality copper cabling. The majority of such tasks are carried out not by specialist fiber optic installers but by existing copper cabling contractors and few, if any, problems are experienced provided that the correct components are selected in the first place. Therefore this chapter does not set out to teach already experienced civil engineers and cabling contractors how to do their job. Only the issues specifically relating to optical fiber are discussed.

The typical installation

Typical is an ambiguous word and a typical fiber optic cabling installation might suggest that the majority of applications proceed in a certain manner with only a few deviating from this path. This would indeed be misleading. The procedure outlined below is an amalgam of the hundreds of installations with which the authors have been associated and to an extent is representative of a typical application without actually mimicking any particular one.

A large installation could comprise long external cabling routes between buildings and shorter routes within buildings. The external routes might be a combination of duct routes, catenary, aerial and wall-mounted sections whilst the internal links might consist of both vertical (riser) cabling together with horizontal (floor) cabling. The installation might require the provision of complete equipment cabinets or may just define the installation of fiber optic termination enclosures into existing cabinets.

The customer may wish to place separate contracts for the civil engineering aspects, where new ducts have to be installed, the cable installation and the final fiber optic works (jointing, testing and commissioning).

The flexibility of this approach is a reflection on the non-specialist nature of all the work with the exception of the final fiber optic content.

The customer may wish to purchase the fiber optic components directly rather than incur the expense of working through the installer. In many cases this is quite acceptable to the installer; however, the customer always runs a greater contractual risk and acceptance testing is vital.

The role of the fiber optic installer can therefore be limited to the termination, testing, jointing, documentation and maintenance of the fiber optic cabling rather than the other non-specialist tasks. As a result the choice of a prime contractor rarely needs to take the transmission technology into account and a commonsense approach should be taken.

Contract management

The introduction to this chapter highlighted a number of contractual interfaces between the installer and the customer and/or third parties. The correct management of these interfaces is vital to guarantee successful completion of the installation (both technically and commercially).

A contractual interface may be defined as a stage at which the responsibility for a product, assembly of products or a service is transferred from one company to another.

Prior to the installation commencing, the various acceptance tests detailed in Chapter 12 should be undertaken by the installer on goods under the control of the installer. When the goods are provided by the customer or third parties then it is rarely acceptable to take the quality of goods or services on trust. In fact to do so may jeopardize the technical or commercial viability of the total installation even though the actions of the organizations may be taken for the best of reasons. Examples of such problems are detailed below.

Supply of fiber optic components by others: fixed cables

Fixed cables may have been pre-purchased by the customer or third parties on behalf of the customer. Alternatively the installer may be asked to extend an already installed system. In these circumstances the installer cannot be held responsible for the condition of the fixed cable but should take steps to ensure that it meets the original specification to which it was purchased in accordance with the relevant testing highlighted in Chapter 12. However, the most important issue is the compatibility of the fixed cable with the components to be supplied by the installer to which it is to be jointed or connected. Incompatibility may be seen as difficulty in jointing or excessive attenuation levels (different kinds of optical fiber may be involved) at joint or demountable connections, and could render the installer liable since the agreed specifications could not be achieved. It is therefore vital for the installer to ensure, as far as possible, that all components are compatible and can be processed in accordance with the specification. When fusion splice techniques are to be used it is wise to ensure that effective jointing can take place and the relevant equipment settings should be established.

Demountable connectors and adaptors

In general demountable connectors achieve optimum performance only when mated with connectors manufactured by the same supplier within the correct adaptor from that supplier. Even if, by chance, improved performance is achieved using a mix of components, the suppliers of the different components are hardly likely to warrant such a combination. As a result it is important for the installer to ensure that all the components supplied by the customer or a third party are to be consistent throughout the installation.

For instance, it has been known for a third party to cut costs by supplying adaptors from a different manufacturer than the connectors supplied on pigtailed, patch or jumper cable assemblies by the installer. When this occurs the installer must highlight, at the earliest possible stage, that the specification may be compromised.

It is also vital to ensure that connector components provided to the installer are complete and that all accessories such as dust caps, washers etc. are available and functional. This can be important since patching fields without dust caps can result in contaminated connector end-faces which may become the installer's responsibility. Absence of the correct washers on adaptors may result in their coming loose from the patching field and potentially affecting the attenuation of the connection made.

Pigtailed, patch or jumper cable assemblies

Where the installer is provided with cable assemblies it is vital to ensure their compatibility with the other cabling components to be supplied. Obvious checks relating to fiber geometry and connector style must be underpinned with acceptance tests for insertion loss against fiber typical of that used within the chosen fixed cable designs. In addition pigtailed cable assemblies should be checked for 'jointability' against the fixed cable, and where fusion splice techniques are to be used the equipment settings should be established.

Termination enclosures

If termination enclosures are to be supplied by the customer or a third party, then it is sensible for the installer to establish that the methods of glanding, strain relief and fiber management are acceptable for the fixed cable designs to be used. Also when the termination enclosure acts as a patch panel it is important to ensure that the panel supplied is suitable for the adaptors to be fitted (in terms of thickness of the panel, the washers and fixing methods supplied).

Cabinets

It is quite common for the termination enclosures to be mounted inside existing cabinets. This is particularly true for 19-inch rack type systems. The installer should ensure that the cable management used within the cabinet is not fouled when the cabinets are closed and locked. This may necessitate the use of recessing brackets. It is normally the installer's responsibility to provide such accessories and their necessity should be established at the earliest possible time.

Documentation

When existing cabling is to be extended it is likely that documentation exists for the installed cabling. Experience has shown that, unless the customer has been fully committed to the upkeep of such information, changes may have been made which are not recorded within that documentation. This is perhaps the most important area in which nothing can be taken on trust and it is the responsibility of the installer to ensure the correctness of the documentation provided by undertaking a sample survey. If for any reason doubts exist, then a full survey must be undertaken prior to the installation commencing.

The implications of accepting faulty documentation can be disastrous and can not only impact just the current installation but can seriously affect the operation of the existing networked services.

Installation programme

The typical installation will comprise procurement of components, undertaking of civil engineering works, cable laying, jointing and commissioning and, finally, documentation of the task performed.

Component procurement

In the majority of cases the critical path item is the fixed cable or cables. Most of the other items to be used within an installation will be comparatively readily available (unless some of the more complex military style connectors are used).

Because there are few standard designs and few standard requirements it is not uncommon for the ideal cable to be unavailable 'off the shelf' and there are two options open to the customer and installer. Either a custom-built design is chosen for which the delivery may be extended or a compromise may be found where a cable of the correct physical parameters is purchased ex-stock, but perhaps the fiber count is in excess of the requirement.

Fixed cable can represent a considerable investment. This factor linked to potential difficulties in procurement places great emphasis on purchasing sufficient to allow for contingencies. Once purchased the value of testing the cable at all contractual interfaces cannot be underestimated.

Whilst not normally considered to be a critical path item the procurement of the various fiber optic cable assemblies must not be overlooked. Cable assemblies should be carefully specified in terms of the cable style, fiber design and connector styles. The latter is particularly relevant for jumper cable assemblies which must be compatible with the connector style adopted on the terminal equipment. As has already been stated a warrantable performance can normally only be guaranteed where all mating components are supplied by a single connector manufacturer and are of an intermatable design. Full insertion loss and, where relevant, return loss measurements shall be recorded with the goods provided. It is also desirable to agree visual inspection standards and, where necessary, environmental testing levels with the manufacturer of the cable assemblies.

Termination enclosures should be assessed for compatibility with the chosen location and when cabinets are to be used to house the termination enclosures, the fixing methods should be established. Also the cable management arrangements outside the enclosures must be checked for compliance with the specification.

Civil engineering works

Frequently undertaken by third party organizations the civil engineering aspects of any cabling infrastructure are rarely influenced by the use of optical fiber. Such works are normally undertaken during the component procurement phase.

Cable laying

It is undeniable that a correctly specified and chosen fiber optic cable can be installed without premium by organizations experienced in the laying of copper communications cables. Nevertheless a sense of apprehension exists due to the understandable, though misplaced, concern that copper conductors must be stronger (and will therefore withstand rougher handling).

It is not always appreciated that high-quality copper communications cables are inherently more complex than their optical counterparts containing glass or silica elements. The performance of a copper cable is frequently a function of the interactions between the various conductors, shielding and insulating materials. For instance, the insulation provided by layers within a copper cable can be drastically altered by poor handling or excessive loading during the laying process. Optical fibers within cables have performance parameters established by the design of the optical fibers. Their performance can only be modified by applied stress rather than changes in their relative position or other physical changes to the cable construction.

Therefore the rules that apply to the laying of fiber optic cables are no more stringent than those for copper. Summarized, these are as shown below:

- *Tensile stress.* The cables shall not be subjected to tensile loads that exceed those specified by the cable manufacturer. For fixed cables being pulled through ductwork this is most effectively guaranteed by the use of mechanical fuses.
- *Bend radii.* There are frequently two separate minimum bend radii specified for a fiber optic cable, one for installation, i.e. under tensile load, and one for operation with the cable under no tensile load. These values should be rigorously complied with.

Provided that the above rules are observed a correctly specified fixed cable can be installed without damage.

Consideration should be given to the quantity of cable left as service loops both within the laid length and at the ends. It is frequently forgotten that the final position of the termination enclosure and the need for access to it makes it necessary to include additional cable at the endpoints. It should also be realized that the need to be able to work upon the fiber inside the termination enclosures necessitates the incorporation of a further contingency at these locations. Finally the probability of fiber damage increases significantly at the cable ends (due to normal wear and

Remote node \ Local node	01A01	02C01	03A01	04A01	04A02	05B01	06A01	07B01
				Cable codes				
01A01			0102				0106	
02C01	0102		0203					
03A01		0203				0305 / 0405/1 / 0405/2		
04A01	Cable codes							
04A02				04A0102 / 0405/1 / 0405/2				
05B01			0305				0506	0507
06A01	0106					0506		
07B01						0507		

Figure 13.1 *Example cable coding system*

tear during handling). Allowing for all these factors it is normal to leave approximately 5 metres of fixed cable when termination enclosures are to be installed. It should not be forgotten that the storage of service loops and the ease with which access is gained to termination enclosures are both very important features of any cabling design.

The end-caps provided with fixed cables are intended to prevent the ingress of moisture and other contaminants into the cable structure. It is important therefore to ensure that the ends of cables are protected during and following installation.

Cable identification is vital. By applying the relevant cable coding system (defined in Chapter 11) it is possible to simply define a specific fixed cable by the use of its destination codes. The nodal matrix shown in Figure 11.2 (page 208) produces a simple cabling coding system shown in Figure 13.1. The use of dual redundant cabling merely requires the addition of a suffix defining the primary and secondary links. It is vital that the organization responsible for the laying of the fixed cable is made aware of the contractual requirements with regard to cable marking and identification.

Termination enclosures and laid-cable acceptance testing of fixed cables

The installed fixed cable must be tested to ensure that no damage has occurred as a result of the installation phase. This is particularly important if the laying of the cable represents a contractual interface.

Once tested the cables should be protected both against subsequent damage and ingress of moisture and other contaminants. This protection is best provided by the termination and jointing enclosures themselves, which

Installation practice 251

Remote node \ Local node	01A01	02C01	03A01	04A01	04A02	05B01	06A01	07A01
			Termination enclosure codes					
01A01		02C01/1					06A01/1	
02C01	01A01/1		03A01/1					
03A01		02C01/2				05B01/1		
04A01					04A02/1	05B01/2		
04A02				04A01/1				
05B01			03A01/2	04A01/1			06A01/2	07A01/1
06A01	01A01/2					05B01/3		
07A01						05B01/4		

Figure 13.2 *Example termination enclosure coding system*

suggests that they should be installed prior to the cabling-laying phase. Although this is not always possible, the early installation of cabinets, wall boxes etc. in their final and agreed locations often limits arguments between the customer, installer and others as to their correct position and orientation. Also the process of moving the various enclosures is much simpler if there are no cables already glanded, tested and spliced into them.

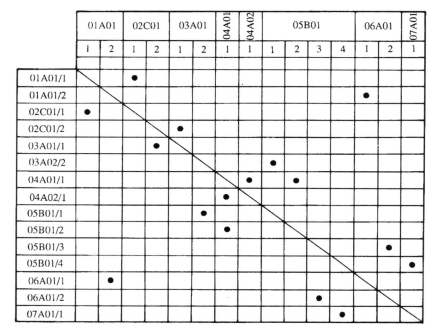

Figure 13.3 *Termination enclosure matrix*

Fiber Optic Cabling

Termination enclosure control sheet			
Issue status			
Local T.E.	Remote T.E.	Cable code	No. of fibres
01A01/1	02C01/1	0102	12
01A01/2	06A01/1	0106	12
02C01/1	01A01/1	0102	12
02C01/2	03A01/1	0203	12
03A01/1	02C01/2	0203	12
03A01/2	05B01/1	0305	12
05B01/3	06A01/2	0506	12
05B01/4	07A01/1	0507	8
06A01/1	01A01/2	0106	12
06A01/2	05B01/3	0506	12
07A01/1	05B01/4	0507	8

Figure 13.4 *Termination enclosure control sheet*

The coding system used for the various nodes produces the nodal matrix mentioned above. The termination enclosures are frequently numbered in accordance with this scheme and an example of this is shown in Figure 13.2 (this follows directly from the nodal matrix defined in Figure 11.2). Using this system each termination enclosure is uniquely identified and the destination of the individual cable from that enclosure is similarly defined using an enclosure matrix as shown in Figure 13.3. The identification scheme allocated to the fixed cables can be extended to the individual fibers within the cables. It is therefore possible to uniquely identify a given fiber element within a given fixed cable within a given termination enclosure. An example of a termination enclosure control sheet is shown in Figure 13.4.

The first part of the laid cable acceptance test is aimed to ensure that the cable to be tested is correctly marked in accordance with the coding system. This may require continuity tests to be performed using visible light sources to ensure that the route taken agrees with the marking attached to the cable itself. This can be performed before any significant cable preparation is undertaken.

When considering long external cable runs, once it has been confirmed that the cable marking is present and correct, the cable must be subjected to the optical performance tests using an optical time domain reflectometer as defined in Chapter 12. The cable should first be prepared by stripping back the various sheath and barrier materials and glanding the cable into the correctly identified termination enclosure. Armouring

Installation practice 253

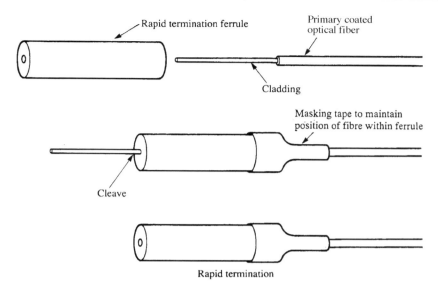

Figure 13.5 *Rapid termination technique*

should be secured to a suitable strength member within the termination enclosure or cabinet. Any other strength members should be secured to the appropriate points and the glands fitted so that approximately 2 metres of optical fiber is left to give the cable installer adequate working lengths. Earthing or isolation of any metal elements within the cables should be undertaken in the manner agreed within the specification.

Once the optical fibers are prepared the OTDR may be used to establish the length of the link and the attenuation of the optical fibers within that link. The unterminated fiber elements must normally be fitted with rapid termination devices (see Figure 13.5) to allow light to be launched from the equipment. The results obtained can be compared with the initial cable tests at the time of delivery and any deviations identified. Localized losses which may have been caused by poor installation technique must be identified and dealt with in the appropriate manner. As each fiber at each cable end is tested it should be marked to allow easy identification during the remainder of the installation process and during any repairs which may be necessary afterwards.

Termination practices

Once the fixed cable has been fully inspected following the cable-laying phase, then the optical fibers within it will be either left unterminated

254 Fiber Optic Cabling

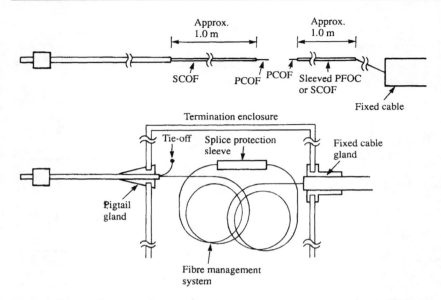

Figure 13.6 *Use of an SROFC pigtailed cable assembly during installation*

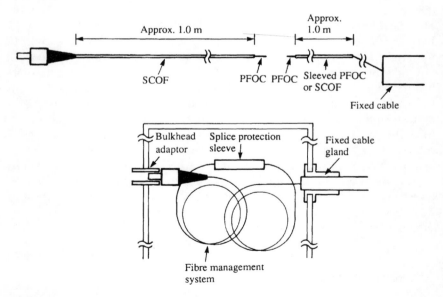

Figure 13.7 *Use of SCOF pigtailed cable assemblies within patched termination enclosures*

(dark), jointed to other fibers (in the case of a joint enclosure or passive node), or connectors will be attached in some way to allow the injection or reception of a signal.

In the case of an unterminated cable then the procedure following laid cable acceptance is quite straightforward. The individual fibers, now uniquely coded and marked, must be coiled and protected from each other, placed in the correct location in the cable management system within the enclosure and the enclosure assembled and fixed to prevent accidental damage by third parties. It should be ensured that the enclosure itself is sealed in an adequate manner to prevent ingress of dust or, where relevant, other contaminants which may be present as advised within the specification.

When jointing to another cable is to take place the laid cable acceptance tests must have been completed on both cables before jointing operations can take place. Once jointing using the chosen mechanical or fusion process has taken place then the fibers must be handled as discussed above and the enclosures sealed and fixed as detailed above.

The options for terminating an optical fiber within a fixed cable which has been fitted into a termination enclosure are either to directly terminate the fiber on-site with an optical connector (multimode only), use a ruggedized pigtailed cable assembly which exits the termination enclosure via a gland and connects directly to another connector or to the terminal equipment or to use a secondary coated pigtail cable assembly which fits into a bulkhead adaptor in the wall of the termination enclosure (thereby creating a patch panel). These preterminated assemblies must be jointed in some way to the fixed cable. The economic decisions surrounding jointing methods have already been discussed in earlier chapters.

When using the ruggedized pigtail approach the pigtail should be stripped back to the secondary coating for a length of approximately 1 metre (see Figure 13.6). The stripped end should be glanded into the termination enclosure and the strength member within the ruggedized cable should be fixed to a tie-off post to provide strain relief. The secondary coated fiber can then be prepared and jointed to the optical fiber within the fixed cable.

When creating a patch panel the pigtailed cable assemblies can be fitted into the rear of the bulkhead adaptors (to protect the terminated fiber end-face) which should be prefitted into the enclosure (see Figure 13.7). The secondary coated fiber can then be prepared and jointed as above.

In either case the final lengths of fiber containing the protected splice mechanism must be coiled, positioned and protected before sealing and fixing the termination enclosure.

In all cases the final highway testing, detailed in the next chapter, cannot take place until the correct fitting of components within the termination enclosure has been completed.

14 Final acceptance testing

Introduction

The preceding chapters have dealt with the theory and design of fiber optic cabling infrastructures. The designs adopted are intended to enable the operation of transmission equipment over cabling which is sufficiently flexible to support the transmission over a considerable period of time.

To validate the design, calculations have been made to prove that the attenuation of the various cabling components and the bandwidth of the individual links are in accordance with the optical power budget of the proposed equipment.

As a result, acceptance criteria have been established against which the installed cabling must be proven. This chapter reviews the possible test method that may be applied to prove compliance with these acceptance criteria.

Not all of the acceptance criteria are optical in nature. Many relate to ensuring that the workmanship has been undertaken to an adequate standard and that the documentation supplied agrees with the actual installed infrastructure.

The inspections and tests undertaken following completion of the cabling installation are generally termed final acceptance testing and represent a significant contractual interface.

General inspection

During the course of the installation it is necessary to make use of the various drawings, diagrams and schematics already referred to in Chapters 11 and 13. These include:

- nodal location diagram;
- nodal matrix;

- block schematic;
- cabling schematic;
- wiring diagram;
- enclosure matrix;
- termination enclosure control sheet.

The purpose of these documents is to guide the installer on site through the entire cabling infrastructure.

The nodal location diagram is simply a list of the nodes to be visited by the cabling. It defines their locations and may list any special access restrictions and contact names, telephone numbers, security codes etc.

As discussed in Chapter 11, the nodal matrix simply defines the interconnection between nodes and can be used to create the cable coding system. It is also normal to include the route lengths, either predicted or actual.

The block schematic is a design document and is not normally required during the installation phase. It forms the link between the nodal matrix and the cabling schematic. The latter details the cables entering each node and defines the termination enclosure into which the cables are fitted.

The wiring diagram is a much more comprehensive document which details the individual fibers and their interconnections. The wiring diagram details all the individual cable codes and termination enclosure codes and defines the coding and marking systems for the fibers themselves.

The enclosure matrix simply links the interconnected termination enclosures and acts as a rapid look-up table.

The termination enclosure control sheets are documents that define the internal configuration of each enclosure. These control sheets are the working documents for the person responsible for the jointing and termination at the various nodes.

General inspection of the installation is made against these documents. The purpose of the general inspection is to ensure that the wiring diagram has been fully complied with and that the standards of workmanship adopted are in line with the specification agreement.

Inspection to prove compliance with the wiring diagram

All termination enclosures should be inspected to ensure that the destinations of the outgoing fixed cables are as per the nodal and enclosure matrices. This is normally carried out at the time of laid cable acceptance testing.

Once the fixed cables have been terminated it is necessary to ensure that the individual fiber elements are correctly routed in accordance with the termination enclosure control sheets.

This is normally carried out once all termination enclosures have been completed, sealed and fitted into their final positions.

It is amazing how frequently the use of very expensive high-technology instrumentation renders the installer seemingly incapable of believing that a basic mistake (such as misidentifying a fiber within a cable) can happen.

A two-man team is allocated to the task, one at each end of the span to be inspected. By launching a strong white light source into each of the terminated fibers within each termination enclosure it is possible to confirm the destination of the individual optical fibers. If there is any deviation from plan it is advisable to find out as early as possible to save delays in overall time scales.

Once it is known that the cabling has been installed in accordance with the wiring diagram the installer can proceed to demonstrate the standards of workmanship.

Inspection to prove standards of workmanship

Obviously when a cabling project involves civil engineering works, cable laying and electrical work (such as earth bonding of cabinets or termination enclosures) the contractual interfaces are also points of inspection.

In general, the standards of workmanship for the non-fiber optic tasks are well established and are not covered here. It is the assessment of fiber optic workmanship which has created problems. The practices adopted during laying fiber optic cables do not differ from those of copper (see Chapter 13) and the main area of inspection must be in the vicinity of the termination enclosures. The key features which must be inspected are as follows:

- Quality of fixed cable strain relief (either as armouring or internal cable strength member).
- Quality of earthing or isolation of conductive cable components (in accordance with the specification agreement). This includes armouring, strength members and metallic moisture barriers.
- Marking and identification of cabinets and termination enclosures in accordance with the enclosure matrix and control sheets.
- Safety labelling of cabinets and enclosures in accordance with national and international standards.
- Marking and identification of fixed cables in accordance with the cable coding system.
- Safety labelling of the fixed cables in accordance with national and international standards.

Secondary issues, but of no less importance, are the quality of the workmanship as it relates to the storage of the service loops of fixed cable near the termination enclosures and the ease with which access can be gained to the enclosures should it be necessary to undertake further work.

These are frequently overlooked until it is too late and should any dispute arise it is far easier to repair or rework the design before the highway becomes operational.

Although normally part of the original technical submission, in response to the customer's operational requirement, the management of any pigtailed, jumper or patch cords (where they form part of the installation) should be inspected. For example, it is frequently necessary to recess the front panels of 19-inch racks within the cabinets to allow the doors to shut without damaging the cables — if this has not been considered then the operation of the highway may become affected.

Finally the customer has the right to inspect the interior of any termination enclosure to establish the quality of the workmanship adopted therein. Key issues are:

- Quality of any strain relief provided internal to the enclosures.
- Quality of the fiber management within the enclosure. Particular attenuation should be paid to accessibility of fiber loops, bend radii of loops and fixing methods used.
- Marking and labelling of the individual elements in accordance with the enclosure control sheet and the wiring diagram.
- Where bulkhead adaptors are used (in patch panel format) their operation should be inspected as should the method and effectiveness of attachment to the panel.
- All dustcaps must be fitted to optical connectors and adaptors.

All the above inspections have little to do with the technical aspects of the optical fiber highway and are aimed to assess the overall quality of the installation. The optical performance of the installed highway is a different matter. It is the sole responsibility of the installer to prove compliance with the optical acceptance criteria defined within the specification agreement. Failure to do so represents a technically based contractual dispute. It is therefore vital to understand the nature of the tests and their applicability before any installation is commenced.

Optical performance testing

The performance and the significance of optical tests on installed cabling is an often misunderstood subject. As has already been discussed with regard to insertion loss measurements, the values of either total cabling loss or individual component losses must always be put into context. However, before any measurement can be made it is vital to define the philosophy behind the testing to be undertaken.

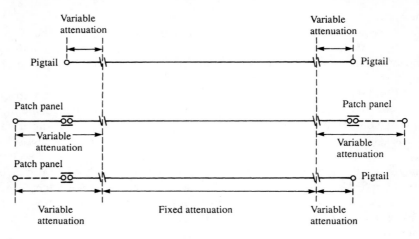

Figure 14.1 *Optical fiber span designs*

At the most basic level, the intervening cabling must have an attenuation lower than the optical power budget specified for the particular equipment. However, this is a very limited criterion since it would be unrealistic to accept a 5 metre jumper cable which exhibited a loss of 15 dB just because the equipment still functioned (with an optical power budget of 19 dB). Not only would it be unrealistic, it would also suggest that a technical fault existed somewhere within that cable which could subsequently fail, rendering the system inoperable.

Figure 14.1 demonstrates the range of possible span designs where a span is defined as a terminated optical fiber cable. The span takes on a pigtailed or patch panel configuration. A number of these spans may be concatenated in a practical system but the basic format of each span will normally comply with these designs. It is not unreasonable then to assess the measurement of installed performance in terms of a fixed content and a variable content.

The fixed content comprises the cable itself plus all permanent joints (in which the alignment between fibers is stable). The variable content corresponds to the demountable connectors or optical joints on the end of the cable. These are treated as variable because of all the potential changes in attenuation produced by the mating of different connectors with those applied to the cabling itself.

The methods used to measure the attenuation of the span designs shown in Figure 14.1 vary and to assess the viability of the measurement it is necessary to review the fixed and variable aspects of the attenuation in each one.

- *Pigtail to pigtail.* The fixed attenuation is the loss due to the optical fiber immediately behind the terminating connectors together with any permanent joints therein. The variable content is provided by the terminating connectors and any variations in launched power due to changes to the alignment of, or basic parameters within, the optical fiber connected to the transmission equipment (or power measurement equipment).
- *Patch panel to patch panel.* The fixed attenuation is the loss due to the optical fiber immediately behind the terminating connectors together with any permanent joints therein. The variable content is provided by the demountable connectors at the patch panels and associated variations in transmitted power due to tolerance within the connectors and basic parametric mismatches within the mated optical fibers at those points (whether under test or in operation).
- *Patch panel to pigtail.* The fixed attenuation is the loss due to the optical fiber immediately behind the terminating connectors together with any permanent joints therein. The variable content is a mixture of the above types.

In all cases it is the variable attenuation which produces the difficulties in measurement. There is no ultimately accurate measurement of an installed optical link because the variable nature of demountable connection losses ensures that unless the operating system is identical in every way to the test system, then the losses seen by the transmission equipment will differ from the test result. In addition, every time the transmission equipment is disconnected, or patch panels reconfigured, the power launched into the cabling by the equipment or the attenuation seen by the equipment will alter.

This makes the assessment of performance somewhat intriguing. It certainly calls into question the issue of repeatability of the measurement and the meaning of the actual measurement. So what does the measurement process achieve?

The measurements made can merely confirm or deny compliance with the original optical specification for a particular span.

However, the first question that one answers is related to the nature of the attenuation which is to be measured. Figure 14.2 illustrates the manner in which the equipment output and input (transmit and receive) powers are specified. The power launched into the chosen fiber from the equipment is defined by the manufacturer as at point A on the diagram. This figure should be modified where necessary in relation to ageing and thermal effects. This is covered more deeply in Chapter 9. Similarly the receiver sensitivity is based around a minimum input power which can, for most purposes, be taken as at point B in the diagram. It makes sense therefore to make a measurement of the performance of the installed span between these two points.

262 Fiber Optic Cabling

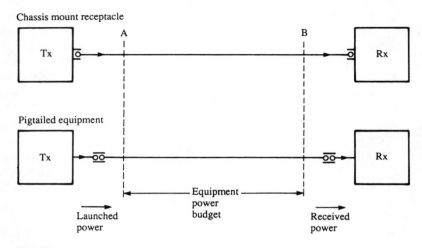

Figure 14.2 *Equipment power budget*

With reference to earlier chapters each terminated optical fiber span has an associated optical loss specification which is determined by the addition of the individual losses of the cable, connectors, joints etc. As a result it is possible to make measurements of a completed optical fiber span in two ways: either by measuring the overall loss or by ensuring that each component in the link meets its own individual specification (thereby achieving the same end).

Overall span attenuation measurement

The use of optical power sources and meters, such as those used to measure insertion loss values for demountable joints, has been the most common of techniques aimed to prove compliance with specification.

The result is a single value which can be compared with a predetermined limit and then recorded within the test documentation.

In this section methods are detailed along with the inevitable measurement errors associated with each method.

Figure 14.1 showed the three fundamental forms of installed cabling. With reference to Figure 14.2 the installed cabling must be compared with the relevant optical loss specification between points A and B (see Figure 14.3).

During training this diagram frequently causes concern and anguish for the trainees. This is because it appears that the end connectors, which are responsible for launching light into the optical fiber and injecting light into the receiver, play no part in the measurement and therefore are not

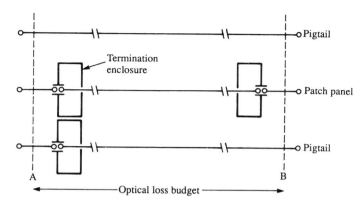

Figure 14.3 *Definition of the cabling attenuation*

proven to be compliant with any specification. This is not so and it is worthwhile to point out why.

The optical power budget of the equipment is defined between points A and B. The minimum launched power at point A is defined for a given fiber geometry assuming that the connection to the equipment is 'good'. Therefore, while the terminating connector must be seen to conform to its own specification (by measurement at the individual component level; whether at the factory, for pre-manufactured terminations, or in the field by other methods discussed below), it plays no part in the assessment of the installed cabling for operation with transmission equipment.

Similarly the received power at the detector is based upon the power available at point B and provided that the terminating connector is 'good' all the power at point B will reach the detector (not entirely true since Fresnel reflection will play a part – this is normally ignored since it is common to all measurement techniques). So yet again, while the terminating connector must be seen to conform to its own specification (by measurement at the individual component level; whether at the factory, for pre-manufactured terminations, or in the field by other methods discussed below), it plays no part in the assessment of the installed cabling for operation with transmission equipment.

This concept can be quite difficult to work with at first but it is nonetheless valid.

Measurement techniques

The use of optical power sources and meters to make a measurement of attenuation between two points in a given link has long been felt to give a more accurate measurement than is achieved by any other method.

264 Fiber Optic Cabling

However, the power source and meter techniques do provide a single-value measurement which can, assuming the measurer has established the correct reference conditions, provide confidence that the installed cabling has met the specification against which it was installed.

Obviously the optical power source must be relevant to the installed cabling specification: the operating wavelength should be in accordance with the proposed operating system and the type of light injection device shall be consistent with the technology adopted. Similarly the optical power meter must match the optical performance of the source. Therefore it is common to measure single mode optical cabling with single mode laser-based sources since that is the dominant operating system, but to use those sources on a multimode 50/125 micron fiber span may give optimistic results which would not be reflected when a multimode system was operated.

Pigtail to pigtail

By reference to Figure 14.3 the measurement to be made is between A and B. Figure 14.4 shows the test method. First a launch reference tail

(a) Setting reference measurement

Optical power source shall
- operate at correct wavelength
- be relevant to operating system (multimode LED, laser, single-mode laser)

Optical power meter shall
- operate at correct wavelength

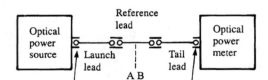

Connectors on test equipment as installed cable

(b) Making the measurement

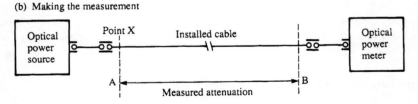

Error: Parametric mismatch between installed cable and reference lead

Figure 14.4 *Measurement of pigtailed-based installations*

lead must be produced. The launch and tail leads should be cladding-mode stripped and core-mode scrambled (as discussed in Chapter 12). The reference lead must be of the same fiber geometry as the installed cabling and be terminated with the same type of connector but must be short enough to introduce no significant attenuation.

The launch, reference and tail leads are connected between the power source and the meter. The resulting power measurement shall be recorded (or the meter 'zeroed') and the reference disconnected.

The reference lead can then be removed from the measurement system and the pigtail-to-pigtail span replaces it. In theory the difference between the reference power and the measurement now recorded is the additional attenuation induced by the intervening cabling between points A and B.

Of course this is not strictly true since there is an error to be addressed. The core diameter and the numerical aperture of the fiber under test may not be identical to those parameters within the reference lead and the results obtained may be higher or lower accordingly. This cannot be assessed readily.

It is necessary to have an agreed measurement tolerance (power meter/light source manufacturers often quote ±0.25 dB accuracy for their equipment) which will vary according to type, style and quality of the equipment used together with connectors used on the pigtails.

Patch panel to patch panel

By reference to Figure 14.3 the measurement to be made is between A and B. Figure 14.5 shows the test method. First, two reference leads must

(a) Setting reference measurement

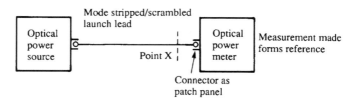

(b) Making the measurement

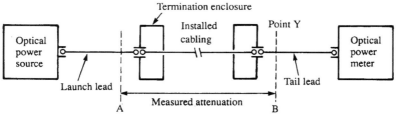

Figure 14.5 *Measurement of patch panel-based installations*

be produced. The reference leads shall be mode stripped (as above) and scrambled.

The reference leads must be of the same fiber geometry as the installed cabling and be terminated with the same type of connector as on the patch panels.

One reference lead is designated the launch lead while the other is termed the tail lead.

The launch lead is connected between the power source and the meter, and the measurement is recorded as the reference value representing the power within the launch lead at point X in Figure 14.5.

Without removing the launch lead from the power source (thereby maintaining the launch condition) the power meter is disconnected and taken to the remote patch panel. The launch lead is connected to the local patch panel and the tail lead is connected to the remote patch panel and to the power meter. The power level now measured represents the power within the tail lead to point Y in Figure 14.5. Therefore the loss between A and B in Figure 14.3 has been measured.

Yet again this is only an approximation since the launch lead and tail leads are unlikely to be the eventual jumper cable assemblies and therefore the loss measured is only representative of the conditions under which the measurement was made.

Patch panel to pigtail

This is an amalgam of the two previous tests.

By reference to Figure 14.3 the measurement to be made is between A and B. Figure 14.6 shows the test method. Only one reference lead is necessary. The reference lead shall be mode stripped (as above) and scrambled.

The reference lead must be of the same fiber geometry as the installed cabling and be terminated with the same type of connector as on the patch panel.

The reference lead is connected between the power source and the meter and the measurement recorded as the reference value representing the power within the reference lead at point X in Figure 14.6.

The power meter is disconnected, taken to the remote pigtailed end and connected to it. The reference lead is connected to the local patch panel and the measurement recorded as the power at point Y in Figure 14.6. Therefore the loss between A and B in Figure 14.3 has been measured.

Similar approximations to those applying to the patch panel measurement above apply to this measurement also.

It is hoped that the reader realizes the very approximate nature of any of the test results. Repeatability of measurements is a major concern and

(a) Setting reference measurement

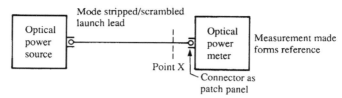

(b) Making the measurement

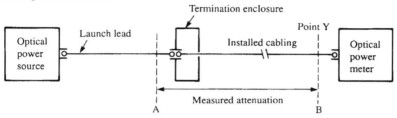

Figure 14.6 *Measurement of hybrid (pigtailed/patch) cabling*

is achievable only if the reference leads are maintained on-site following the testing.

The single value measurement produced by power source and power meter methods is able to give confidence that the overall link performance is within specification but can allow individual non-compliant components to remain undetected. Single ended testing can mask faults at connectors. It is possible for a chipped connector to launch light into a detector and appear satisfactory, but when light is injected into that same chipped connector, the angled end-face can reflect a large quantity of light and render the link unusable. Double ended testing is therefore recommended for total security.

A power meter/light source is the most appropriate method for acceptance testing of multimode links up to 2 kilometres and single mode links up to 1 kilometre. A power metre/light source will show if there is a problem, but not where or what that problem is. For faultfinding and characterizing long installed lengths of fiber then an OTDR, optical time domain reflectometer is required.

Optical time domain reflectometer testing of installed spans

From the discussion of single-value measurements of attenuation of installed optical fiber spans it has been seen that the practical methods of identifying losses between points A and B in Figure 14.3 have certain limitations.

To totally validate the performance of installed components and the techniques used to interconnect them it is necessary to undertake the assessment of localized losses such as joints, demountable connector joints etc.

By adopting professional levels of quality assurance the performance of purchased goods may be assessed. This has been discussed in Chapter 12. It is nevertheless important to prove that the individual components and their methods of installation meet their individual specifications following installation (and prior to contractual handover to the customer). The only effective method for analysing long cable lengths, or faulty short lengths, is to use an optical time domain reflectometer.

The OTDR chosen must operate at the specified wavelength and must be capable of performing valid measurements on the length of cable installed. Equipment is now available which can characterize lengths as short as 20 m in all operating windows, and this section is illustrated with actual traces from such equipment. This may seem an obvious point but the authors have seen totally useless traces taken using very expensive yet unsuitable kit, whereas a much lower investment matched by skilled testing personnel would have produced all that could be desired.

The basic operating principles of OTDR measurement were explained in Chapter 12 and are not complex.

With reference to Figure 14.3, the three types of installed cabling are described as pigtail to pigtail, patch panel to patch panel and patch panel to pigtail. The measurement technique for all is basically identical since it is the individual component losses which are being addressed rather than the overall loss between points A and B. If all the component losses are within or conform to specification then, by definition, the overall attenuation will also lie at, or within, specification.

Before attending on-site a pair of test leads must be produced. One can be called a launch lead whilst the other functions as a tail lead, but they must conform to the following requirements:

- They should be manufactured using the same design of optical fiber used in the link to be characterized.
- They should be terminated at one end with a connector suitable for connection to the OTDR and at the other end with a connector of the type as that fitted to the pigtail or patch panel.
- They should be long enough to allow the launch loss of the OTDR to be dissipated (further discussion of length takes place below) prior to the end of the cable.
- They should be mode stripped and scrambled.

The launch lead should be connected to the OTDR and its length established using the refractive index supplied with the installed cable.

Final acceptance testing 269

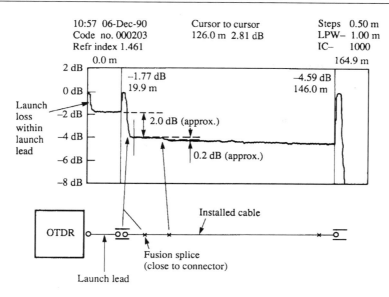

Figure 14.7 *Single ended measurement of an installed span using an OTDR*

If this is not available then a nominal figure such as 1.484 should be used.

When the launch lead is connected to the installed span then a trace of the form shown in Figure 14.7 will be produced. It allows measurement

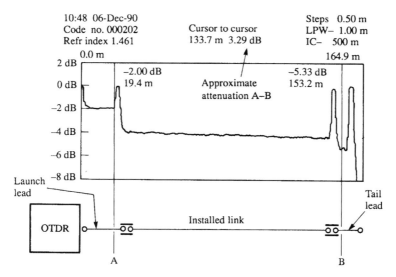

Figure 14.8 *Double ended measurement of an installed span using an OTDR*

of the local mated connector pair, any joints (assuming the resolution is adequate) and the installed cable. The attachment of the tail lead to the far end of the cabling allows measurement of the remote connector pair. This is shown in Figure 14.8.

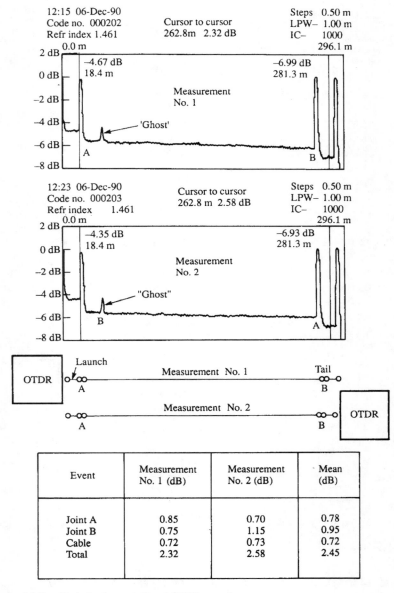

Figure 14.9 *Detailed analysis of OTDR results*

Final acceptance testing 271

To be pedantic, the OTDR measurement should be undertaken in both directions when a measurement is being attempted. This requires the launch lead and tail lead to be left connected to the cabling whilst the OTDR is transported from one end of the span to the other. The individual results for each component should then be averaged and this figure quoted as the correct value. This is demonstrated in Figure 14.9. However, this can be time consuming and frequently single ended measurements without the use of tail leads are accepted without significant error.

The arguments for double ended measurements are quite strong since the measured loss across a particular event can be different when measured from opposite ends of the cable. This is due to a combination of effects which become significant when the fibers on either side of a joint are from different manufacturers etc. However, single ended measurements are common if the purpose of the testing is to either confirm or deny compliance with the specification and not to make an absolute measurement of a particular component within the cabling.

The problems of 'ghosting' discussed below can be made significantly worse if a tail lead is used.

Choice of launch leads and tail leads

One of the most confusing effects of the use of OTDR equipment results from the appearance of 'ghost' reflections. Figure 14.9 demonstrates the effect.

The ghost is produced by a second (or third or more) reflection from a given event. It is most easily observed when characterizing a span using a launch lead. Light injected by the OTDR is reflected back from the first demountable joint and returns to the OTDR. Fresnel reflection occurs at the end of the launch lead connected to the OTDR and some light travels back to the first demountable joint, which in turn experiences reflection and travels back to the OTDR, and so on.

There are at least two ways of removing ghosting effects. One is to reduce the level of reflection at the first demountable joint. This can be done by inserting index-matching fluid between the ferrule ends. Unfortunately this has two major disadvantages. The first is that it reduces the measured insertion loss of the connector pair by up to 0.35 dB (by removal of Fresnel loss) and is therefore an unethical method of altering the results produced against a fixed specification. Also the connectors must be carefully cleaned afterwards. This may be difficult but, if not done fully and correctly, can create contamination areas which attract dirt and fungal growths.

The second way is to use launch leads which are longer than the links to be tested. This may at first seem ridiculous when networks can easily be extended beyond a kilometre in total length. However, the cost of primary coated optical fiber is low enough to render the concept commercially viable for certain key installations.

272 Fiber Optic Cabling

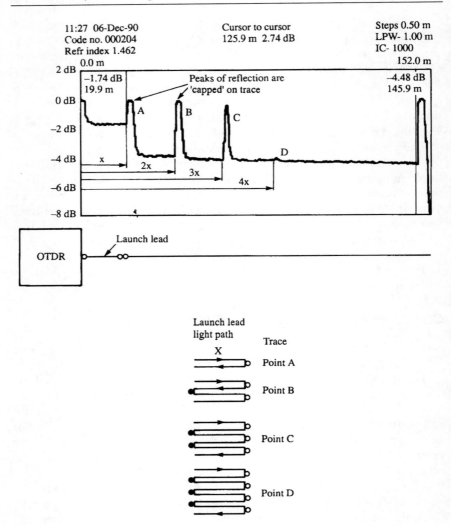

Figure 14.10 *Ghosting and its analysis*

Most professional installers have a range of standard lengths for launch leads which must be chosen to minimize the confusing impact of ghosts on the traces.

Figure 14.10 shows an extreme case of ghosting and explains how to interpret a complex trace. The obvious methods of detecting ghost traces lie in their periodicity and also the fact that they actually have no impact upon the loss at the points where they appear.

Comparison between test methods and the results obtained

As has already been stated that the methods of testing installed spans can rarely be accomplished in a definitive fashion and the results obtained can only confirm or deny compliance with the agreed specification.

However, it is worthwhile reviewing the actual losses measured for identical events using different methods and equipment.

The most important issue in the measurement of any optical loss is the distribution of light in the launch lead. To some extent this is obvious. For instance, if a short multimode launch lead is used and the power source is a single mode laser, then the light emitted from the launch lead will have a very low numerical aperture. This will make any joint into which that launch lead is connected look much better than if an LED source was used which explored the full modal distribution of the launch lead fiber.

This straightforward example suggests that commonsense should also prevail and that the measurement conditions should be consistent with the operational use of the cabling. However, the situation is considerably more subtle than it first appears.

A fair and even-handed measurement necessitates a launch condition which features mode stripping and scrambling. Mode stripping has already been discussed and ensures that no light is carried in the optical cladding. Mode scrambling aims to ensure that the light carried down the launch lead explores the full modal distribution available within the fiber. This is normally achieved by introducing a tight mandrel wrap but can be achieved by using a long length of optical fiber (which can be used to eliminate 'ghosting' on the OTDR traces).

Using such a launch lead the results of a given event will be identical (within measurement tolerance) whether the source is a laser or an LED. Failure to use such a lead can introduce errors which can deceive the installer and customer alike.

Differences of up to 0.5 dB can be introduced at demountable joints by lack of attention to such conditioning of the light with the launch lead. This difference can worsen when the core diameter and numerical aperture of the launch lead increase.

15 Documentation

Introduction

The professionally installed cabling infrastructures being specified for campus and backbone highways are frequently expected to provide services over a considerable period of time and are rarely static. Changes will be made to both the cabling configuration and the transmission equipment. As a result there is an overwhelming need to document the installation correctly and in a manner that will allow updates to be accommodated easily.

Contract documentation

Each installation is different and therefore the specifications and other contract documents will differ also. It is impossible to be dogmatic about what should be included in the contract documentation. However, there is a wide variety of documents which may be needed such as:

- operational requirement;
- design proposal;
- technical specification;
- contractual specification;
- invitation to tender;
- bill of materials (initial);
- tender submission;
- quality plan;
- certificates covering staged completion of the contract;
- change notes or variations to contract;
- final specification;
- bill of materials (final);
- certificates of conformance (for materials);

- acceptance test certificates and test results;
- laid cable test results;
- final test results;
- final system documentation.

Once the cabling is completed, and paid for, the administrative aspects become less important and are generally filed away. The remainder of the documents relate to the installed infrastructure and should be considered 'live' and subject to change. These are:

- operational requirement (to remind the customer and the installer of the original concept);
- laid cable acceptance test results (to provide a performance baseline for currently unterminated fibers within the cabling);
- final acceptance test results (to provide a performance baseline for currently terminated fibers within the cabling);
- final system documentation (to fully define connectivity);
- final system specification (to set down performance requirements for future modifications).

These documents can be viewed as the technical documents covering a flexible structure rather than the contractual documents covering the initial installation. Future installations will build upon the technical documents whilst the individual contracts may differ significantly.

Technical documentation

The purpose of technical documentation is to enable the customer, installer and other third party organizations to easily identify routes, cables, termination enclosures and to provide a performance baseline against which any further work can be assessed. Additionally it should facilitate repair and maintenance functions (see Chapter 16). All of the above requirements should be achieved under the overall guidance of the operational requirement, i.e. the original document produced by the customer which forms part of the specification agreement at each installation or modification phase.

The final system documentation supplied to the customer by the installer encompasses all the various aspects of the technical documentation and is detailed below.

Final system documentation

During the design phase a number of documents were produced. In order for the installation to proceed a nodal location matrix and nodal matrix had to be produced. To ensure the correct level of connectivity a block

276 *Fiber Optic Cabling*

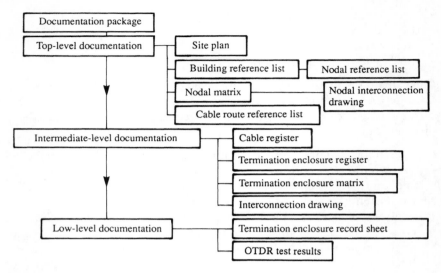

Figure 15.1 *Documentation package structure*

schematic, cabling schematic and a wiring diagram were generated which in turn produced the enclosure matrix and individual termination enclosure control sheets.

These documents are discussed in Chapter 14 and examples are shown in Chapter 17.

It is important for the final system documentation to include these documents together with the final test results. Equally important, however, is the manner in which they are incorporated. A documentation structure should be developed that allows the cabling modifications to be encompassed within the final system documentation in a controlled way that minimizes the effort involved.

A tree-and-branch structure is suggested that features the concept of Issue Status for each document. A possible structure is shown in Figure 15.1.

A change to a cabling span may require changes to all or only some of the layers of the documentation. However, an update, or up-issue, to a document at one level will necessitate an upgrade of the next level and so on.

Nevertheless this approach does have basic flaws if not handled correctly. As was suggested above, one must be careful that the smallest alteration at one of the bottom layers does not necessitate the modification and up-issue of *all* the layers above.

To prevent this it is sensible to construct the entire installation documentation using three layers, between which the documentation links are simplified and minimal. This approach involves the consideration of a cabling system at three management levels.

The first is confined to assessing the infrastructure at the overall interconnection level by defining the connectivity between the major nodes, thereby describing the cable route information. This represents the most simplistic level to which changes are only made by the addition or deletion of cable routes.

The second delves deeper into the architecture by apportioning the components within each cable route. This provides information relating to the individual routes by defining the coding for termination enclosures and the cables themselves. This level needs to be modified only when changes are made to individual cable routes and/or the enclosures into which they are terminated.

The final level forms the physical connectivity layer. This details the manner in which the individual cable routes are handled within the individual enclosures. These documents are modified when any small changes are made to the connectivity.

The benefit of this approach is that a minor change, such as the addition of a pair of pigtails to both ends of an existing cable route, does not need to affect the documentation beyond the lowest, or third, level. Since such changes are much more likely than those at the top level it makes sense to minimize the task involved in documentation upgrade.

Contents and layout

Having outlined the justification for, and a possible approach to, the use of a structured documentation package for a structured cabling project, it is necessary to identify the nature of the actual documents needed.

Top-level nodal information

The key is to formalize all the documents included. This means that the use of subjective items should be avoided wherever possible. The type of documents to be included may vary but must be designed and incorporated with upgrade and modification in mind. Some suggestions are:

- *Site plan.*
- *Building and nodal coding system and reference list* (see Figure 15.2). This gives each building a numeric code, each floor an alphabetic code and each node a numeric code within that building and floor. This allows the reader to immediately link the accepted building description etc. with a code that should be used throughout the entire documentation pack. This should be designed to allow easy expansion or contraction of the installed cabling.

278 Fiber Optic Cabling

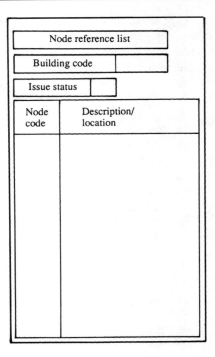

Figure 15.2 *Building and node reference list formats*

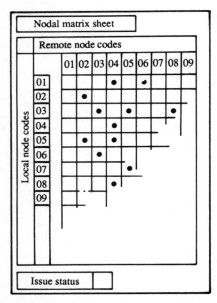
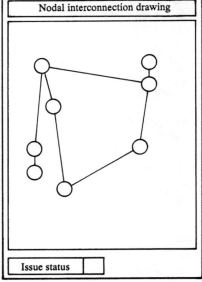

Figure 15.3 *Nodal matrix sheet and interconnection drawing*

- *Nodal matrix and nodal interconnection drawings* (see Figure 15.3). These can be produced at any number of sensible levels. Perhaps an inter-building matrix can be produced together with an inter-floor and an inter-node matrix. In this way the addition of nodes on an individual floor will affect only the relevant matrix and not the level above.
- *Cable route coding* (see Figure 15.4). This defines the form of cable route coding system adopted between each of the nodes. This is not to be confused with the actual codes applied to the individual cables running on the routes since certain routes may involve more than one cable between two nodes. Descriptions of the components are included in the next level of documentation.

Intermediate level nodal information

For each nodal interconnection defined in the top-level documentation the following information is supplied:

- *Cable design, type and coding register* (see Figure 15.5). This references the style of each cable together with the coding applied to it.
- *Termination enclosure register* (see Figure 15.6). This references the style and location of each enclosure together with the coding applied to it.
- *Termination enclosure matrix and interconnection drawings* (see Figure 15.7). These give termination enclosure details and define the coding applied to those enclosures at either end of the cable. The drawings should

Cable route reference list

Node A	Node B	Cable route code	No. of cables

Issue status	

Figure 15.4 *Cable route code reference list*

280 Fiber Optic Cabling

show termination enclosure interconnection on a closed-loop basis. This is better described in Chapter 17 where a case study is presented. This latter approach makes modifications to individual nodal interconnections much simpler to deal with.

Cable register				
Route code	Cable code	Cable description		
		Fiber geometry	No. of elements	Physical
Issue status				

Figure 15.5 *Cable register*

Termination enclosure register				
T.E. code	Location	Style	Cable codes	Remote T.E. code
Issue status				

Figure 15.6 *Termination enclosure register*

Documentation 281

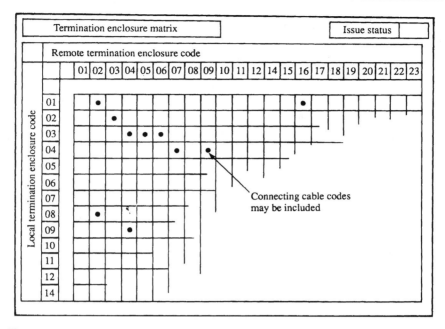

Figure 15.7 *Termination enclosure matrix*

Low-level connectivity information

At this level, documentation relates to individual cables and the methods by which they are interfaced to the nodes at the termination enclosures. Since this is a true physical layer it is not surprising that the documentation supplied is both the most detailed and the most technical, and that it contains all the test documentation relating to the physical highway in a given fixed configuration. The following information is supplied:

- *Termination enclosure record sheets* (see Figure 15.8). These show the actual connections to the cables at both ends of a cable. They enable the reader to identify the specific span within a cable and its destination by virtue of its coding either at a patch panel or as a pigtail.
- *Optical time domain reflectometer* traces of individual optical fibers (see Figure 15.9).
- Other test result information, e.g. power meter test results.

Specification register

The intermediate and low-level documentation packages make reference to the specification of cables, termination enclosures and termination

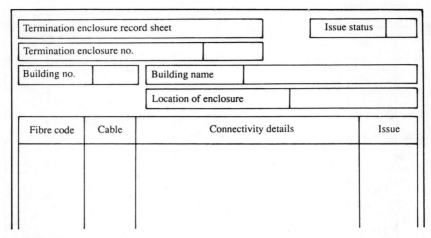

Figure 15.8 *Termination enclosure record sheet*

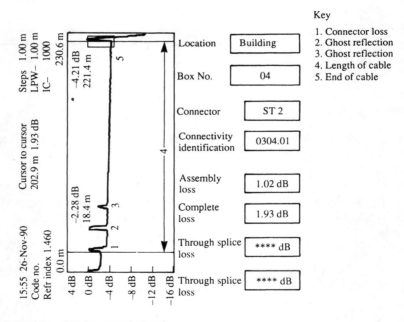

Figure 15.9 *OTDR record sheet*

methods (such as connector types or splice mechanisms). All relevant specifications and drawings must be included in the specification register.

Each of the above document or sets of documents must be given an issue status and a unique identifier. This allows subsequent changes to be made by the removal of certain documents and their replacement with upgraded versions by the customer in a controlled manner under the guidance of the installer or customer.

The function of final highway documentation

The value of good-quality documentation for larger installations is immeasurable. It defines connectivity, it provides confirmation that specifications have been met and it supplies information that could be vital in case of emergency by rapid identification of all interconnection at termination enclosures etc. Additionally it provides a statement of performance against which future measurements can be compared to identify deterioration.

However, documentation is only as good as the last upgrade and if it is not controlled it will rapidly become useless. It may even become a source of contractual dispute if reliance is placed upon wrong or historical information which does not reflect actual connectivity. It is therefore important to produce documentation in bite-sized pieces which taken together form a cohesive picture and which are designed for update.

International standards concerning project documentation

EN 50174

EN 50174-1 *Information technology – cabling installation – Part 1: Specification and quality assurance*, offers a systematic way to control documentation throughout the life of an I.T. cabling system, both for copper and optical cabling.

Quality plan

A quality plan is proposed for the project containing:

- Cabling component acceptance: acceptance test methods and inspection criteria.
- Installation competence.
- Inspection.
- Documentation of installed cabling.
- Identifiers.
- Repair and maintenance philosophy.

The documentation

The documentation should include:
- The installation specification.

- The quality plan.
- Evidence of conformance of the components.
- Cable acceptance test records.
- Cable assembly test records.
- Delivery information, e.g. date of receipt, batch numbers etc.

Final cabling documentation

The final cabling documentation includes:

- Site plans including all nodes, pathways, cables etc.
- As-built drawings.
- Evidence of conformance to the specification.
- Handover certification.
- Other information as required.

The documentation can be in paper or electronic form or within a custom-made software cabling database of which there are many on the market today.

Administration

EN 50174 describes how to administer the cable system, either in paper or electronic format, to maintain the value of that cabling system over time.

- *Identifiers.* A unique code that distinguishes an item or location within the cabling system.
- *Labels.* Labels are fixed to the component to carry the identifier and possibly other information.
- *Records.* The collection of information relating to the cabling system.
- *Work orders.* A document (paper or electronic) that requests and records changes to the cabling system.

Identifiers

The items that need to be identified are:

- *Pathways.* Each pathway should have a unique identifier. The records should then show where the pathway goes, what type it is and where the earthing points are.
- *Spaces.* Spaces are the equipment rooms, telecommunications closets, entrance facilities etc.
- *Cables.*
- *Termination points.*
- *Earthing and bonding points.*

TIA/EIA-606 Administration Standard for the Telecommunications Infrastructure of Commercial Buildings

TIA/EIA 606 considers the following areas of a cabling project that have to be administered:

- Terminations for the telecommunications media located in work areas, telecommunications closets, equipment rooms and entrance facilities.
- Telecommunications media between terminations.
- Pathways between the terminations that contain the media.
- Spaces where terminations are located.
- Bonding/grounding, as it applies to telecommunications.

ISO/IEC 14763 Information technology – Implementation and operation of customer premises cabling

ISO 14763 contains three parts:

ISO/IEC 14763-1	Administration
ISO/IEC/TR3 14763-2	Planning and installation
ISO/IEC/TR3 14763-3	Testing of optical fiber cabling

Parts 1, 2 and 3 are relevant to fiber optic cabling installations and much the same advice is given about identifiers, labelling, pathways etc. as in EN 50174 and TIA/EIA 606.

16 Repair and maintenance

Introduction

The concept of providing some level of repair and maintenance service for a cabling medium is quite alien to those involved in copper solutions. Nevertheless the importance of the fiber optic cabling structure and the communications services operating on it together with the long life expectancy make it a candidate for some level of after-sales support.

This chapter reviews the relevant issues and the options open to the customer.

Repair

The bandwidth and the low attenuation levels offered by optical fiber have extended the signalling rate and physical extent of communications networks respectively. Failure of the communication path is therefore more unacceptable since alternatives are not always available. This often makes the customer nervous and the idea of fast response repair contracts is not unreasonable, at least at first sight.

Nevertheless it should not be forgotten that the fastest response to failure is good design. If a cabling infrastructure is well designed the problems associated with cabling component or equipment failure will already have been considered and the fault analysis and action plan will be predefined.

So whilst the customer is within his, or her, rights to ask for a repair contract from the cabling installer, or from a third party, it is incumbent upon all concerned that a sensible approach be adopted to the installed cabling and the equipment connected to it. This approach should combine design aspects and practical fault analysis issues. The design elements have already been discussed in earlier chapters and include spare optical fibers and even dual redundant cable routes. Additionally, highly modular constructions can be used allowing easy replacement of damaged cabling sections (particularly useful in high-connectivity infrastructures). At the equipment level either hardware or software reconfiguration can be implemented to reroute information over alternative cable routes.

The following section reviews fault analysis techniques that are aimed at initiating self-help amongst customers, thereby reducing the need for expensive fast response repair contracts which may never be used.

Fault analysis techniques

There can only be two reasons for the catastrophic failure of a given communications link. Either the equipment has failed (at one end or both) or the cabling has become in some way defective. Software-based surveillance of a communications network can normally pinpoint a failure to a particular receiver within a transceiver unit. The failure is characterized by the inability of the receiver to pass on to the user the signal traffic. This can only be for one or more of the following reasons:

- malfunction within the receiver (misreading an adequate power input level signal);
- malfunction within the cabling (thereby limiting the optical power reaching the receiver);
- malfunction within the transmitter responsible for injecting the light into the optical fiber (either by corrupting an otherwise acceptable signal or by injecting an insufficient optical input level).

To identify which failure mechanism applies requires nothing but commonsense and the ability to resist the urge to panic and start changing every component in sight in an attempt to return the system to its former operating condition.

The three options do not include the possibility of bandwidth restrictions which would necessarily corrupt the data being transmitted whilst not affecting the optical power received at the transceiver. However, bandwidth is not believed to deteriorate under normal operating conditions.

The normal way in which it becomes known that a fault exists depends upon the terminal equipment registering a 'low light' signal which is highlighted by network management software or by physical observation. This 'low light' indication can be used to effectively fault-find to the degree necessary to allocate responsibility to either the cabling or the equipment.

The first step is to identify if the entire cabling link has failed in some widespread and catastrophic manner. This is achieved by assessing whether a 'low light' condition exists at both ends of the duplex link. If it does then it is very likely that the fixed cable is damaged in some way and must be tested accordingly with an optical time domain reflectometer.

288 Fiber Optic Cabling

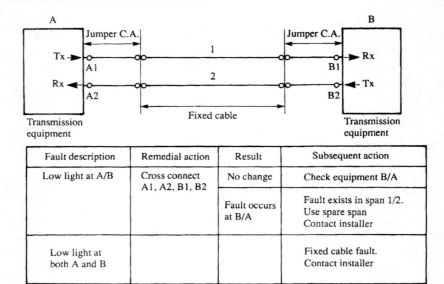

Figure 16.1 *Fault analysis techniques*

If the 'low light' condition affects only one end of the system, then it is necessary to isolate the fault. With reference to Figure 16.1 the connections to the terminal equipment should be reversed (at both ends) to see if the fault follows the optical fiber or is constant and therefore likely to be equipment related.

If the fault is traced to a particular transmitter or receiver then there is no alternative but to change the equipment. If it is felt that the fault lies within a given optical fiber path, assuming the entire cable is not damaged, then the installer should be contacted to effect a repair.

Nevertheless there are further steps that can be taken by the customer which may allow earlier repair. If the optical fiber path includes jumper or patch cable assemblies then a controlled substitution may be undertaken in an attempt to locate the faulty element, termination or joint.

These basic fault analysis procedures are valid for most networks and, if used correctly, can save a great deal of embarrassment and money on behalf of the customer.

Repair contracts and their contents

The benefit of good design is that the need to repair is minimized by the use of replaceable items such as jumper or patch cables. However, the unexpected takes place eventually and the capacity to achieve an effective, speedy and permanent repair is largely dependent upon planning for the unexpected. Repair contracts should be seriously considered. The contents of such contracts vary with the type, size and importance of the installation and the communications services offered. The best and most comprehensive contracts are produced following an analysis of all the potential fault locations and the contract will include all materials to cover the repair task in each location type.

It is sensible to include in such a package spare components of the types used in the initial installations (such as fixed cables, termination enclosures, jumper and patch cords). Rapid access to these components in an emergency can make all the difference between a minor setback and a major catastrophe. But it is not always sufficient to merely keep a spares holding of the components used originally. It is worthwhile reviewing whether or not there are additional components or pieces of equipment which may be necessary to complete a repair.

For instance, where all the fixed cables run through totally water-filled ducts it makes sense to ensure the provision of waterproof enclosures should a cable break have to be dealt with.

Maintenance

Maintenance contracts

The concept of cabling maintenance is rather new but is rapidly gaining acceptance as the operating requirements for high-speed data highways are being extended.

Maintenance contracts can be offered separately from, or in conjunction with, repair contracts. The primary purpose of maintenance contracts is to enable changes in cabling performance to be identified, quantified and, if necessary, remedied before any lasting damage or catastrophic failure can occur within the cabling infrastructure. To undertake this work a performance baseline must exist – this underlines the need for full and comprehensive documentation as defined in Chapter 15.

When such contracts are already operating it is normal for an annual check to be made, usually on a sample basis to prevent disruption to the communications networks operating on the highway. The checks will pay particular attention to the condition of connector end-faces, cabling loss characteristics (from an optical time domain reflectometer) and the general mechanical well-being of the installation.

Customer-based maintenance

When faced with all the benefits of optical fiber it is easy to overlook the one big drawback. Optical fiber interfaces do not like dirt.

Customers should seek guidance from the installer on the correct cleaning and general maintenance procedures relating to the installation for which they will become responsible.

Summary

A fiber optic cabling project is not finished once it is installed – indeed it is only just beginning to perform its primary function. Consideration should be given to the maintenance of the structure and also, since everything fails eventually, its repair.

17 Case study

Introduction

This chapter is a form of case study since it is based upon a number of individual installations undertaken by a British installer. As a result, the names of the buildings etc. are totally fictitious but the calculations, design decisions, specifications and documentation packages are based upon real installations for real customers.

Network requirements

The network was to be installed at an industrial plant. The installation was to be considered in three separate phases, the first being campus or inter-building cabling (thereby interconnecting existing copper networks in each building), the second being backbone cabling within certain of the buildings and the last phase being the provision of optical fiber-to-desk locations in other buildings.

The dimensions of the plant were relatively large with some communication links in excess of 2 km. Nevertheless the majority of links were less than 700 metres in length.

The services to be transmitted on this network were to be Ethernet (to IEEE 802.3 100BASE-FX and 1000BASE-SX) with eventual migration to ten gigabit Ethernet between buildings. Other specific point-to-point services were to be introduced as required.

The initial plant was one of a number owned by the customer and it was desired to produce a design and implementation strategy which covered all sites.

Preliminary ideas

When faced with the task of determining an overall strategy the following aspects must be addressed:

- What design implications are dictated by current requirements?
- What design implications are dictated by future requirements?
- If the implications are divergent, how can they be married together?

The first issue to be addressed was that of a common nomenclature to be used for all buildings, floors and nodes. This had to be able to be extended to meet future requirements without the need for change to the underlying structure.

The next requirement was to analyse the defined communications requirements for the current installation. IEEE 802.3 100BASE-FX and 1000BASE-SX both have associated specifications for both attenuation and bandwidth which would have implications for fiber geometry, termination enclosure design etc.

It was possible that the types of point-to-point services to be considered would place unrealistic requirements on the fiber geometries necessary to meet all Ethernet designs. Alternative geometries had to be considered, in the form of composite cable designs.

Solutions had to be generated for inter-building, intra-building and fiber-to-the-desk environments and a range of products were to be specified which could be used throughout the system.

For all installations, both present and future, a set of specifications was to be generated for the optical performance of the components used together with the joints produced.

An agreed test method specification was to be produced.

Finally a documentation package was to be formulated which would allow the non-expert customer to find his or her way through the installed system on any of the company's sites.

Initial implementation for inter-building cabling

Figure 17.1 shows the initial implementation.

There were three major communications sites:

- Penistone Building
- Howes Court
- Joseph Centre

There were three major spurs:

- from Penistone to Sharpley Tower, Kler Block and Lee Block;
- from Howes Court to Buckley and Barrowcliffe Buildings;
- from Joseph Centre to Corker Blocks Nos 1 and 2.

This set of ten buildings represents the total initial implementation. There were two further unnamed buildings which might require connection at

Case study 293

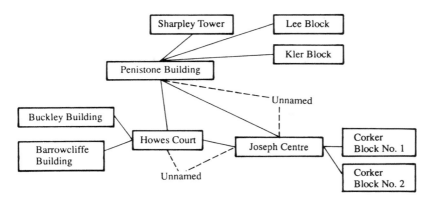

Figure 17.1 *Case study: initial implementation*

some later date. With these exceptions all further expansion of the fiber optic cabling infrastructure would be internal to buildings.

It was proposed that each of the customer's sites would be described in the following manner.

Nomenclature

Each distinct building location was given a two-digit numeric code. The number of floors in that building was established and each floor was given an alphabetic code (starting at the lowest level with A). This allowed the most basic coding, of nodes i.e. ends of fixed cabling structures.

This system allowed up to 100 buildings, each having 26 floors, to be connected per site. This was felt to be more than adequate for most purposes and was accepted by the customer.

For the initial implementation it was decided to allocate the following codes as seen in Table 17.1.

This most basic proposal, having nothing to do with fiber optics or communications in general, assists greatly during the design phase of the task.

Interconnection requirements

Figure 17.1 shows geographically that there were three distinct groups of locations. The three prime locations, coded 01, 05 and 08, were to be connected directly, forming a ring type structure. This ring structure could be utilized by intelligent, redundancy-based Ethernet switches, by token-ring protocols, ATM or channel systems such as ESCON or Fiber Channel for Storage Area Networks. The key issue is the resilience of the ring to main cable failure.

Table 17.1 *Building coding scheme*

Building code	Description	Floors	Floor code	Node
01	Penistone Building	Basement	A	01
		Ground	B	—
		First	C	—
		Second	D	—
02	Sharpley Tower	Basement	A	—
		Ground	B	01
		First	C	—
		Second	D	—
		Third	E	—
		Fourth	F	—
		Fifth	G	—
		Sixth	H	—
03	Kler Block	Ground	A	01
		First	B	—
04	Lee Block	Ground	A	01
		First	B	—
05	Joseph Centre	Basement	A	—
		Ground	B	01
		First	C	—
06	Corker No. 1	Ground	A	01
07	Corker No. 2	Ground	A	01
08	Howes Court	Basement	A	01
		Ground	B	—
		First	C	—
09	Barrowcliffe Building	Basement	A	01
		Ground	B	—
		First	C	—
10	Buckley Building	Basement	A	01
		Ground	B	—
		First	C	—
11	Unnamed No. 1	Ground	A	—

Connection of locations 02, 03 and 04 to location 01 (and similarly 06 and 07 to 05, 09 and 10 to 08) was required to be undertaken in a star format (to suit Ethernet switches in each of the connecting buildings); however, migration from Ethernet to other protocols had to be possible using the installed cabling infrastructure. These spur runs were felt to be less critical and there is no redundant path if the main cable fails –

however, only the remote site would suffer and the customer had decided that this is a risk that could be taken.

Connection of the unnamed locations, 11 and 12, was initially not necessary but would be considered if relevant.

Installed ducts and civil works review

Figure 17.2 shows the existing duct routes across the site. The installed base was considerable and there was sufficient duct space available to suggest that any significant civil engineering works would not be necessary.

However, a number of issues were raised:

- The cable runs did not follow the shortest paths between buildings. This is typical and the decision had to be made whether it was cheaper to buy more cable than to dig more ducts.
- There were two points on the duct drawing where the cable had to cross water by the use of an existing catenary or messenger wire. This was a potential opportunity for lightning strike and might have had implications for the fixed cable design.
- Existing ducts passed within 50 m of locations 11 and 12. It was agreed that if it proved necessary to use these locations cable could be introduced to the locations during the initial phase.

Figure 17.3 shows the proposed cable routes and their lengths. These lengths were, in most cases, of little relevance; however, there were a number of long runs which were reviewed in terms of the traffic they were intended to carry.

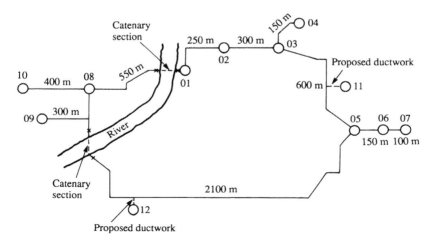

Figure 17.2 *Existing and proposed ductwork*

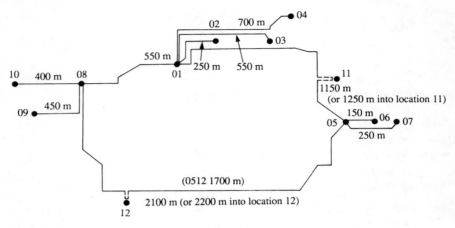

Figure 17.3 Cable routes and lengths

Table 17.2 *Optical LAN operation by power budget*

Network application	Optical power budget ISO 11801 2nd edition in normal type, *TIA/EIA-568-B.1 in italics*			
	Multimode		Single mode	
	850 nm	1300 nm	1310 nm	
10BASE-FL,FB	12.5 (6.8)	*12.5* *(7.8)*		
100BASE-FX		11.0 (6.0)	*11.0* *(6.3)*	
1000BASE-SX	2.6	*3.2* *(3.9)*		
1000BASE-LX		2.35	*4.0* *(3.5)*	5.0	*4.7*
Token Ring 4,16	13.0 (8.0)	*13.0* *(8.3)*		
ATM 155	7.2	10.0 (5.3)	7.0	*7.0 to 12.0*
ATM 622	4.0	6.0 (2.0)	7.0	*7.0 to 12.0*
Fiber Channel 1062	4.0		6.0	*6.0 to 14.0*
FDDI		11.0 (6.0)	*11.0* *(6.3)*	10.0	
The figures in parentheses are for 50/125 performance				

Table 17.3 *Optical LAN operation, by maximum supportable distance*

Network application	Maximum supportable distance ISO 11801 2nd edition in normal type, *TIA/EIA-568-B.1 in italics*					
	Multimode				Single mode	
	850 nm		1300 nm		1310 nm	
10BASE-FL,FB	2000 (1514)	*2000* *(2000)*				
100BASE-FX			2000 (2000)	*2000* *(2000)*		
1000BASE-SX	275 (550)	*220* *(550)*				
1000BASE-LX			550 550	*550* *(550)*	2000	*5000*
100BASE-SX★	300 (300)					
Token ring 4,16	2000 (1571)	*2000* *(2000)*				
ATM 155	1000 (1000)	*1000* *(1000)*	2000 (2000)	*2000* *(2000)*	2000	*15000*
ATM 622	300 (300)	*300* *(300)*	500 (330)	*500* *(500)*	2000	*15000*
Fiber Channel 1062	300 (500)	*300* *(500)*			2000	*10000*
FDDI			2000 (2000)	*2000* *(2000)*	2000	*40000*

The figures in parentheses are for 50/125 performance, otherwise 62.5/125 fiber
★100BASE-SX is described in TIA/EIA-785

Fixed cabling design

The customer wished to operate equipment, bought independently, which meets the requirements of 100 Mb/s, 1000 Mb/s and 10 000 Mb/s Ethernet.

Table 17.2 shows the relevant optical parameters for spans necessary to facilitate the connection of such equipment.

The maximum distances achievable are as detailed in Table 17.3.

Final location interconnection matrix

Because of the relatively short distance between the installed duct routes and the unnamed buildings 11 and 12 it was decided to install short duct routes into these locations from the main route. However, it was originally decided to run cable into these buildings and straight out again leaving a service loop without actually invading the fixed cabling structure. Thus these locations would not, as far as the connectivity documentation is concerned, visit either location.

The other decision that had to be made resulted from Figure 17.3 where the main cable route between locations 01 and 05 passed locations 02 and 03. There was an argument to suggest that only one cable should be installed and that each location should be visited allowing both direct connection to that location and daisy-chain connection to the next. However, this was not pursued because the cost savings (on cable and cabling installation) would not be significant when balanced against the need for changes in cable design (for that part of the network), the additional potential for cabling failure (on the main ring), the deviation from a simple and consistent design and the additional cost of jointing and testing the spliced-through cable. Obviously if the length of combined route had been greater this argument would have been weaker.

For similar reasons it was decided that the two spur runs from location 05 to 06 and 07 should be separately cabled.

As a result Figure 17.4 shows the final cabling routes. The location of the primary nodes in each building had been defined and the nodal matrix shown in Figure 17.5 was prepared.

Cable route coding

External cabling was coded in a simple manner. All routes were defined by a four-digit numeric code, the first two digits being the source location and the last two being the destination location. However, the source is defined as the lower of the two codes.

Cable route	Node location	Length (m)
01-02	01A01 02B01	250
01-03	01A01 03A01	550
01-04	01A01 04A01	700
01-05	01A01 05B01	1250
01-08	01A01 08A01	550
05-06	05B01 06A01	150
05-07	05B01 07A01	250
05-08	05B01 08A01	2200
08-09	08A01 09A01	450
08-10	08A01 10A01	400

Figure 17.4 Final cabling routes

Case study

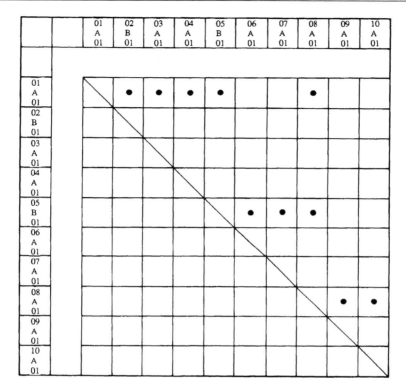

Figure 17.5 *Nodal matrix*

The following gives a full listing of the routes.

Route	Description
0102	Existing duct, water filled, minor civil engineering into building
0103	Existing duct, water filled, minor civil engineering into building
0104	Existing duct, water filled, minor civil engineering into building
0105	Existing duct, water filled, minor civil engineering into building. Duct required to location 11
0108	Existing duct, water filled, minor civil engineering into building. Portion of route crosses river via catenary
0506	Existing duct, water filled, minor civil engineering into building
0507	Existing duct, water filled, minor civil engineering into building
0508	Existing duct, water filled, minor civil engineering into building. Duct required to location 12. Portion of route crosses river via catenary
0809	Existing duct, water filled, minor civil engineering into building
0810	Existing duct, water filled, minor civil engineering into building

Materials choice

Fixed cable construction – external

The total length of external fixed cable required was approximately 6750 metres as measured between the nodes. Allowing for service loops used in the installation process the total length might be forced towards 7000 metres.

The basic requirement to service both Ethernet and other protocols did not define the quantity of optical fibers within the cables used. First, it was necessary to define the method by which the Ethernet network was to be implemented. For instance, if it were decided to place Ethernet switches at locations 01, 05 and 08, then it would really only be necessary to include two elements in each cabled route. If, however, it were decided to use only one repeater at location 01, 05 or 08, then certain of the cable routes would need the inclusion of eight elements.

To provide sufficient capacity for all such configurations and to allow additional point-to-point services it was decided to implement the entire site infrastructure with a cable containing 24 optical fibers.

To progress towards 10GbE (ten gigabit Ethernet) any length over 300 metres must be implemented in single mode fiber. For gigabit Ethernet, any route over 550 metres must be implemented in single mode. 50/125 (OM2) fiber is appropriate for links up to 550 metres for gigabit Ethernet, but 10GbE needs laser enhanced OM3 style of fiber for up to 300 metres (although the relatively expensive wavelength division multiplexing can be used on OM2 50/125) but at the time of the project design, OM3, although proposed, was not readily commercially available.

Blown fiber was considered. However, the route of 2100 metres was beyond the current capability of blown fiber and the customer believed the provision of 24 fibers would be sufficient for the lifetime of the current infrastructure.

Conventional optical cable was decided upon with 16 OM2 50/125 fibers and eight ITU-T G652 single mode fibers.

Two cables could be installed, one with 16 50/125 fibers and one with eight single mode fibers. One composite cable was specified, however, for the following reasons:

- Two cables take two operations to install and take up twice as much duct space.
- No distributor could be found that stocked 7 km of eight-fiber single mode cable, so it would have to be made to order.
- Due to the quantity, manufacturers were happy to quote for a composite cable that met the specific requirements of the project, and the cost was less than two separate cables. Note that manufacturers are often

happy to quote for special designs if the demand is in multiples of 2 km, because that is how they buy the fiber.

The cable would be based upon a loose tube construction containing six tubes. Two tubes contained eight, individually coloured, primary coated 50/125 fibers and a third contained eight single mode fibers. Three tubes were solid plastic and were there to keep the cable design round. The three populated tubes were gel-filled as a final water barrier. The remainder of the construction featured a non-metallic GRP (glass reinforced plastic) central strength member, water-blocking tapes and threads to act as a moisture barrier and a polyethylene oversheath. A discussion arose regarding the use of a full gel-fill; however, it was decided that, as no cables were terminated in a drawpit, the need for a gel-fill was not overwhelming.

The relatively short lengths of aerial installation across the two water courses were discussed and it was agreed that the most cost-effective solution would be to standardize on the above cable as it was metal-free.

Fixed cable construction – internal

The nodes defined in the initial implementation were intended to be the primary activation point for each building and as such they lay just inside the buildings at a convenient cable termination point. It was perceived that the transmission equipment would be positioned at these points, thereby providing the interface between the copper internal cabling structure of the building and the optical highway. These primary nodes were intended to support secondary nodes as the optical fiber was extended into the buildings and it was necessary to have a strategy for the interiors of the buildings even though no internal fixed cable was required in the initial plan.

There were two possible degrees to which this could be achieved. The first involved the extension of the external fixed cable into the building whereas the second was considered to be the provision of a comprehensive optical infrastructure within the building independent of the external connectivity (with the proviso that the two had to be connected at the primary node being installed at the outset).

To service the first requirement it was decided to use standard internal cable designs where possible, since the lengths of the internal cable routes would not justify a full custom design. Proposals were made demonstrating the penetration of optical cabling to each floor via the use of a standard premises distribution cable consisting of 12 tight-buffered elements. 62.5/125 fiber would have been adequate but it was decided to stick with 50/125 throughout so that any internal cable could be patched onto the campus backbone cable if this was ever required. The 50/125 actually worked out cheaper anyway.

The bundle of 12 fibers would be wrapped with aramid yarn, acting as an impact resisting layer, and sheathed with a material with low fire hazard, zero halogen material. It was not thought necessary to consider the penetration of the single mode optics into the building. The design of the termination enclosures required an external–internal joint at the primary node.

With regard to the widespread introduction of the optical medium into the building it was decided to carefully evaluate the possibilities of utilizing a blown fiber solution.

Connector choice and termination enclosures

It was decided to use the SC duplex connector as the multimode system connector, i.e. at all termination enclosures, and it was agreed that the single mode SC connector would be used for single mode applications.

For the external fixed cable it was necessary to use a 1U high 19-inch rack as the basis for the termination enclosures. The front panel contained three fields of four SC duplex adapters (i.e. terminating 24 fibers). Two fields were for multimode, which were coloured beige to denote multimode fiber. The single mode adapters were blue to denote non-angled physical contact single mode connectors.

Since the termination enclosures might eventually be directly connected to internal termination enclosures (and thence to internal fixed cables) it was decided to specify the enclosure for use as either a pigtailed or a patch panel variant.

The rear panel of the subrack contained a single gland suitable for the incoming fixed cable (12.5 mm diameter) and a number of predrilled, blanked holes to accommodate any future cables needing to enter the panel. Large tie-off posts were fitted into the baseplate to accept the central strength member of the fixed cables. Note that the metal content within the cables, if any, would have to be earthed at a point as soon as possible after entering the building. All external cables should always be terminated within 5 metres of entering a building anyway.

The ports on the front panel of the termination enclosure were numbered as follows:

Multimode   01 to 16
Single mode A to H

It was decided to fit all termination enclosures with brackets allowing them to be recessed into the cabinet to which they were eventually fitted. This prevents damage to the jumper or patch cable assemblies connected to the front panel during the opening and shutting of cabinet doors or movement of equipment within the cabinet.

A coding system had been defined for the termination enclosures which was based upon the node coding system with a sequential suffix.

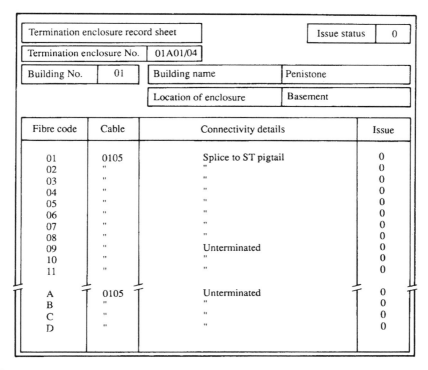

Figure 17.6 *Termination enclosure record sheet*

This system is used to formulate the 'bill of materials'. This allowed blank termination enclosure record sheets to be drawn up for all nodes which form the basis of the installation working instructions, see Figure 17.6.

Cable assembly specification

As it had been agreed to terminate the installed cables by the fusion splicing of preterminated pigtailed cable assemblies a specification was drawn up defining test methods, conditions and acceptance parameters for all preterminated assemblies (pigtailed, patch or jumper cable).

- SC connectors, multimode
  specification IEC 60874-19-1 and IEC 61754-4
- SC connectors, single mode
  specification IEC 60874-14-5 and IEC 61754-4
- SC adapters, multimode
  specification IEC 60874-19-3
- SC adapters, single mode
  specification IEC 60874-19-2

- The random-mated insertion loss across any two connectors must not exceed 0.75 dB.
- Single mode connectors and adapters must be visually identified by the use of the colour blue.
- Multimode connectors and adapters must be visually identified by the use of the colour beige.
- Patch cables must be of different colours for single mode and multimode (preferably blue and beige) and the fiber type must be inkjet-printed onto the cable sheath.
- Each patchcord or pigtail must be individually wrapped or packaged.
- The return loss across any pair of randomly mated multimode connectors shall be better than 20 dB, for single mode connectors the figure shall be 35 dB.
- Splices shall be permanent fusion splices protected with a heatshrink sleeve. The maximum insertion loss across any splice shall be 0.3 dB.
- Any optical connector may be visually inspected and rejected if physical damage extends across the core of the fiber.

Bill of materials (fiber optic content)

It was necessary to produce a bill of materials (BOM) for the initial implementation and also to allow future expansion of the infrastructure to be undertaken using a common set of piece parts and specifications. The BOM includes both material and labour aspects. The materials content of the initial implementation was considered first.

Jointing requirements of the initial implementation

The details of the jointing to be undertaken at individual termination enclosures were agreed which allows the compilation of a BOM covering the materials needed.

Bill of materials (components)

The BOM relating to the fiber optic portion of the installation is detailed below:

Item	Description	Quantity
1	Fiber optic cable:	7000 m

Construction: Non-metallic, multiple-loose tube external grade optical cable suitable for immersion in water-filled cable ducts. Two tubes must each contain eight individually coloured, primary coated multimode fibers and one tube must contain eight individually coloured, primary coated single mode fibers. The tubes will be sufficiently colour coded to allow unique identification of every single fiber in the cable.

Specification:

- Diameter   13 mm max
- Finish/sheath   black polyethylene
- Temperature range   −20 to +70°C
- Tensile strength   1500 N min
- Minimum bend radius   175 mm
- Crush resistance   500 N/cm
- Water penetration   IEC 60794-1

Delivery and packing requirement: Cables will be supplied as three times 2-kilometres and one times 1-kilometre on wooden drums with all exposed cable ends protected with waterproof heatshrink covers. The cable will be printed on the sheath, in white, 'Optical fiber cable – external use only. Year of manufacture'.

Optical parameters:

- Sixteen off 50/125 μm and eight single mode primary coated optical fibers.
- Unless defined elsewhere, the optical fiber specification shall be according to IEC 60793-2 and more specifically:
  50/125 overfilled launch bandwidth at 850 nm;   500 MHz.km
          overfilled launch bandwidth at 1300 nm;   500 MHz.km
          attenuation at 850 nm;   3.5 dB/km max
          attenuation at 1300 nm;   1.5 dB/km max
- Single mode fiber shall be according to ITU-T G652.

2   Patch panel termination enclosure:

- 1U 19-inch rack, recessed by 100 mm.
- Finish to black painted steel.
- Front panel to be loaded with eight multimode, beige, SC duplex adapters to IEC 60874-19-3 and four, blue, SC duplex adapters to IEC 60874-19-2.
- The back of the panel shall have suitable cable glands to accept and support at least two optical cables in the diameter range 10 to 15 mm.
- The interior of the panel shall contain fiber management equipment to manage at least 24 1-metre pigtails including up to 24 fusion splice protector sleeves. The fiber minimum bend radius, as defined by the cable manufacturer for single mode fiber, will not be infringed at any time.

3   Pigtail cable assembly:

- 750 to 1000 mm of primary coated fiber to the same specification and colour code as the main cable and supplied by the same manufacturer.

- Terminated at one end with SC multimode optical connectors to IEC 60874-19-1 and IEC 61754-4 or SC single mode connectors to IEC 60874-14-5 and IEC 61754-4.
- All pigtails to be factory made and tested and individually packaged.

4   Fiber optic splice protection sleeves to any major telecommunication standard, e.g. Bellcore, British Telecom etc.

5   Patch cable assembly:

- Two metres of tight-buffered (900 micron) optical fiber to the same specification as the main cable and supplied by the same manufacturer.
- Patch cable to be 2.8 mm (nominal) diameter with low flammability sheath (IEC 60332-1) and aramid yarn strength members.
- Terminated at both ends with SC multimode duplex optical connectors to IEC 60874-19-1 and IEC 61754-4 or SC single mode duplex connectors to IEC 60874-14-5 and IEC 61754-4.
- All patchcords to be factory made and tested and individually packaged.
- Patch cables must be of different colours for single mode and multi-mode (preferably blue and beige) and the fiber type must be inkjet-printed onto the cable sheath.
- Duplex patchcords shall be cross-over, i.e. A to B and B to A.

Bill of materials (labour)

In order to define the labour content of the BOM it was necessary to define the requirement for testing. The overall task to be undertaken was as detailed below:

- Civil engineering works (preparation of existing ducts, provision of new ducts, supply and installation of traywork/conduit within buildings).
- Acceptance testing of fixed cable.
- Installation of termination enclosures.
- Cable laying (including all necessary marking and provision of service loops).
- Supply of other materials.
- Preparation of fixed cable.
- Jointing of pigtailed cable assemblies.
- Final acceptance testing.

A quality plan was prepared to support the above tasks in line with EN 50174. It is not reproduced in full but included the following points:

- *Initial cable acceptance.* The cable was required to be delivered on four drums (3 × 2000 and 1 × 1000 m) and it was agreed that the installer

should witness testing of the cable at the place of manufacture. It was required that each of the optical elements in each of the reeled cables should be subjected to OTDR testing (multimode 850 nm and 1300 nm, single mode 1310 nm only) and that the results obtained should represent the performance baseline. The drums were to be delivered direct to site and subjected to a physical examination (to identify any shipment damage).
- *Pre-installation acceptance.* As the installation was intended to take place over some three months (due to external factors) it was stated that the cabling contractor, employed by the customer to lay the cable, desired contractual confidence that the cable was fully functional immediately prior to being installed. Part of the quality plan provided for sample testing of the optical fibers on each drum as it became available for use. The OTDR results at this stage were to be compared with the performance baseline. It was decided that testing four of the single mode elements at 1310 nm would represent an adequate sample.
- *Final acceptance testing.* For campus cabling projects it is usually unnecessary to do OTDR tests for acceptance testing. Due to the long lengths of cable in this particular project, and the presence of single mode fiber, the customer decided that he wanted both OTDR and power meter tests. The OTDR tests were at 1300 nm for the multimode fiber and 1310 for the single mode, in one direction only. The test specification clearly called up the following:
  - copies of training certificates for the OTDR operators were required in the response to tender;
  - pulse width settings must be according to the OTDR manufacturer's instructions for cable plant of this length;
  - launch and tail cables of the length appropriate for the cables tested must be used according to the OTDR manufacturer's instructions for cable plant of this length.

Hard copies of the OTDR traces must be supplied as part of the final documentation. It is a good idea for somebody who understands OTDR traces to have a look at them and formally accept. Too often ODR traces are called for, but as nobody understands what they are looking at they are just filed away.

Power meter testing was to be done on all links, at 850 and 1300 nm for the multimode and 1310 for the single mode, in both directions. The results must be presented on a test sheet as shown in Figure 17.7. Note in this chart that there is a column for calculating the expected attenuation values according to ISO 11801 rules. The four columns on the right are the actual measurements taken at both wavelengths (multimode) and in both directions. If all of the measured values are equal to or less than the calculated values, then the system has passed the test and can be accepted.

Another common fault of optical cable testing is to simply record lists of attenuation readings. Without referral to what attenuation is expected for that link, a simple list of attenuation values is meaningless. The power meter and light source must be correctly calibrated according to the manufacturer's instructions. The recent addition of optical power meter add-ons to

OPTICAL TEST REPORT   SHEET REF. NO. .... DATE ....

INSTALLATION CO...
END-USER   ....
SITE ADDRESS ....
               ....
               ....

CABLE I/D: ....
TUBE COLOUR OR NUMBER FROM REFERENCE TUBE: ....

CABLE LENGTH: .... NO. OF FIBRES IN TUBE...................

FIBRE TYPE: .... NO. OF FIBRES IN CABLE: ....

CABLE TYPE/PART NO: ....
PANEL TYPE/PART NO: ....
CONNECTOR TYPE/PART NO: ....

NUMBER OF SPLICES IN LINK ....
NUMBER OF CONNECTORS IN LINK ....

WAVELENGTHS TESTED ....
MAKE & MODEL OF TEST EQUIPMENT ....

FIBRE NO.	FIBRE COLOUR	CALCULATED ATTENUATION	MEASURED ATTENUATION			
		850/1300 nm	A to B		B to A	
			850 nm	1300 nm	850 nm	1300 nm
1						
2						
3						
4						
5						
6						
7						
8						
9						
10						
11						
12						

TESTER NAME ....   SIGNATURE ....   DATE ....

Figure 17.7 *Power meter test sheet*

hand-held copper testers has been a mixed blessing. A simple power meter/light source combination will give a simple figure for attenuation and it is up to the installer to decide if that figure is acceptable. A fiber optic power meter 'add-on', however, wants to make a pass/fail decision, as it does for the copper tests. To do this it sends a pulse of light and measures the round trip travel time from the reflected pulse, thus it can calculate the distance (within the limits of its own internal accuracy and its estimate of what the refractive index is). From the distance, and its programmed knowledge of ISO 11801 parameters, it can judge if the measured attenuation is acceptable or not. To make this judgement, however, it needs to know how many connectors are in the line because it will add or subtract 0.75 dB for every connector (and 0.3 dB for every splice it knows about). The machine therefore needs to know exactly how many connectors there are, or else it will start making assumptions, and an assumption of 0.75 dB at 1300 nm is the same as adding or subtracting 750 metres of fiber! It is essential therefore that fiber tester 'add-ons' are calibrated exactly according to their manufacturer's instructions. Unfortunately there have already been many projects where large test regimes have been completely wasted due to meaningless test reports from these types of testers.

This agreement upon the form and quantity of testing enabled a BOM to be produced for the labour content of the fiber optic cabling installation.

Both BOM lists having been produced together with a clear appreciation of the tasks involved, it was possible to progress to the next stage of installation planning.

Installation planning

It was necessary to create installation planning documentation and this was based upon modified versions of the termination enclosure record sheets as shown in Figure 17.6. It details the task to be undertaken at each termination enclosure including all mechanical information, jointing information and test requirements.

Summary

This chapter has taken the reader through a typical installation from the design stage to the final stage of detailed installation planning. Obviously it is impossible to recreate a blow-by-blow account of the actual installation but it should be realized that if the planning and design has been controlled appropriately then the actual installation will normally run smoothly (unless influenced by outside factors). Following the installation phase comes the documentation package and this has been handled in some detail in Chapter 15.

18 Future developments

Introduction

A time traveller from 1990, who worked with optical campus cabling and was able to jump forward to 2001, would easily recognize most of the products in use today. We still use the same types of 50/125, 62.5/125 and single mode fiber, albeit with a better performance and a lower price. The optical connectors we use today are very different from the SMA and biconnic connectors of 1990, but still easily recognizable as ferrule-based optical connectors.

The often-heralded trend of fiber-to-the-desk is still on a very slow growth path due to the combination of high-quality twisted pair copper cabling and cheap, microprocessor-based, digital signal processing chips that can insert sophisticated coding techniques onto copper cable to extract every last usable hertz of bandwidth.

Optical fiber started as a telecommunications technology and that is where 97% of the value still remains, so it is to the world of telecoms we have to look to see what will be the influential factors in the LAN/premises cabling environment in the coming decade.

Exotic lasers

Datacommunications has traditionally used LEDs because they are cheap and can drive 155 Mb/s over 2 kilometres on multimode fiber. Lasers are traditionally very expensive and LEDs are very cheap. However, single mode fiber is approximately one-third the price of multimode fiber, and so if lasers ever came down in price then the whole rationale for multimode fiber might disappear.

VCSELs (vertical cavity surface emitting lasers) seem to be the key. 850 nm VCSELs hit the market in the late 1990s for gigabit Ethernet applications. Making the jump to 1300 nm single mode VCSELs has proven to be much harder, however. In 2001, prototype samples of 1300 nm, 2.5 Gb/s VCSELs, based on indium phosphide, finally became available for early trials. When 10 Gb/s, 1300 nm, single mode VCSELs become available for prices measured in the tens of dollars, it will spell the beginning of the end for multimode fiber.

Other products still in the laboratory are electrically powered organic lasers. An organic laser could be cheaper to mass produce than today's gallium arsenide devices. LEDs may still make a comeback, however. Researchers at Leuven in Belgium and the University of Erlangen in Germany have achieved 1.2 Gb/s transmission speed with an LED based on thin-film gallium arsenide and aluminium gallium arsenide with a surface textured with microscopic pillars.

Lasers have always been made to work at one wavelength. In the world of wavelength division multiplexing it would be an ideal to have a laser that could be tuned just like an electrical oscillator. This would require extremely specialized optics but research is being carried out to achieve this. One method is discrete single-frequency tuning, using a Fabry-Perot interferometer, to give specific resonances locked to 50 GHz intervals, known as the ITU grid.

Another method is to use quantum dots. These are tiny areas of light-emitting semiconductor that are so small that the wavelength emitted is directly proportional to the size of the 'dot'.

New optical fibers

Over the last decade the only changes in data communications fibers have been the disappearance of large-core fibers, such as 100/140, the slow acceptance of single mode, the non-appearance of plastic fiber and the probable introduction of a new laser-grade, high-bandwidth 50/125.

The telecommunications market has seen the arrival of dispersion shifted fiber, i.e. optimized to work at 1550 nm, then non-zero dispersion shifted (NZDS) fiber optimized for wavelength division multiplexing, larger core area single mode to lower the power density (and hence to lower intermodulation effects) and dispersion compensating fiber to make up for the dispersion reintroduced by the NZDS fiber. Future developments would seem to centre on the perfection of all these fibers.

One, slightly more bizarre, development is 'holey' fiber. This is fiber with a core made up of a lattice-like array of hexagonal, and other shaped, air spaces, which act as light guides. These fibers may have many applications including dispersion compensation and wavelength tuning.

Instead of optical fibers, mirrored waveguides may be used. Tubes made out of perfect mirrors can direct light around tighter bends than optical fibers and so may be useful for miniaturized optical components. The mirrored waveguide also transmits light without changing the polarization states; essential for long distance transmission.

Plastic fiber is still in the lab. Deuterated poly methyl methacrylate has shown that it can give a bandwidth and attenuation performance close to multimode fiber; but until fibers such as these can show either a price or performance superiority over silica fibers it is difficult to see the motivation in turning to them.

Next generation components

There are many optical components around today which have been developed to maximize distance and bandwidth capabilities of optical fiber, such as Bragg gratings, Raman amplifiers and a host of other esoteric devices. Wavelength division multiplexing, WDM, remains the best way of optimizing the bandwidth inherent in single mode optical fiber, although with one version of ten gigabit Ethernet we have seen the first proposed use of WDM in a datacoms environment.

The current state of the art with off-the-shelf equipment is 40 channels of 10 Gb/s data, i.e. 400 Gb/s per fiber. This is with a WDM spacing of 100 GHz, or about 0.25 of a nanometre. Improving upon this requires more selective filters and wider bandwidth amplifiers. New WDMs based on arrayed waveguide gratings (AWGs) will offer 80 channels of 40 Gb/s at 50 GHz spacing, or 3.2 Tb/s (terabits per second). The technology is capable of going to 160 channels, also at 40 Gb/s per channel, or 6.4 Tb/s per fiber. Up to 800 channels could be possible giving each single mode fiber a potential bandwidth of 32 Tb/s.

Over what kind of distance this could be achieved is more problematic, as polarization mode dispersion will limit the capacity in practice, but multi-terabit capacity over tens of kilometres is certainly achievable.

Optical switching is another major area of research. At present switching and other data manipulation is done by converting the optical signals back to electrical signals, processing them, and then converting them back to optical signals again. This method slows down the overall network and potentially adds noise and errors. Optical amplifiers can now purely amplify the signal whilst remaining completely in the optical domain, but now all-optical switching is starting to make an appearance.

A range of technologies is under scrutiny, including microelectromechanical systems (mercifully abbreviated to MEMS), liquid crystals, electro-optic, thermo-optic, acousto-optic and 'bubble' switches which combine inkjet printer technology with planar lightwave circuits.

MEMS switches consist of arrays of tiny mirrors sitting on a silicon chip. The mirror can be tilted using electromechanical means to deflect a laser beam input into a choice of output fibers. MEMS can switch inputs and outputs within a few milliseconds, i.e. hundreds of times per second. This is adequate for network reconfiguration but not full packet switching which needs to be done billions of times per second.

A MEMS switch is protocol transparent, that is, it can switch any kind of optical data stream, regardless of what's in it. However, that also means that it cannot read the address destination of a packet, so this would have to be entered into the switch by another means.

New coding techniques

Traditionally, there are three ways to modulate information onto an electromagnetic wave; these are by amplitude, by frequency or by phase, or by any combination of all three. Copper cabling systems currently use extremely complex combinations of modulation techniques to maximize information content. Optical systems are usually relatively simple and just use the digital version of amplitude modulation, called PCM, pulse code modulation.

On top of the basic modulation technique, digital signals also need to be encoded to give information about clocking rates and to prevent long streams of 'ones' and 'zeros' desynchronizing the receiver. Optical systems generally use the basic NRZ (Non-return-to-zero) coding method but higher bit rates use RZ (return-to-zero).

The available bandwidth may be shared out amongst various data streams by way of time division multiplexing (TDM) or frequency division multiplexing (FDM). The wavelength division multiplexing used in the optical fiber industry is just another form of frequency division multiplexing (wavelength equals velocity of light divided by the frequency, with the velocity in turn dictated by the refractive index). Wavelength division multiplexing, WDM, is often referred to as dense wavelength division multiplexing or DWDM in a telecommunications context of 0.25 nanometre spacings to differentiate it from wide or coarse WDM which may have 25 nanometre spacings.

A different approach, although related to WDM, is OCDMA, or optical code division multiple access. OCDMA uses a broadband light source that is then selectively filtered. Unique selections of the filtered spectra are then encoded with data. The selected spectra are often likened to a bar code. Unlike FDM or WDM there are no multiplexers or demultiplexers involved and so the system should be cheaper. All channels are present at all points of the network, so users can quickly reconfigure a system into ring, star or mesh topologies. Thirty channel, 2.5 Gb/s systems appeared as prototypes in 2001 and may offer a lower cost, high bit-rate solution for both telecommunications and local area networks.

Appendix A Attenuation within optical fiber: its measurement

Fiber optic systems rarely rely upon the measurement of absolute optical power. Rather they are designed, defined and measured in terms of power ratios that are presented in the form of decibels (dB).

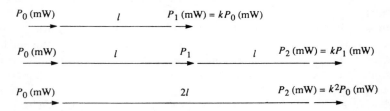

Figure A.1 *Concatenation of optical power*

To illustrate, the example shown in Figure A.1 is used in which a given length l of fiber is used as a reference. An absolute optical power P_0 (mW) launched into it is attenuated and an absolute output power of P_1 (mW) is detected at the far end, where:

$$\frac{\text{detected power}}{\text{launched power}} = \frac{P_1 \text{ (mW)}}{P_0 \text{ (mW)}} = k$$

If a further length l of identical fiber is invisibly jointed to the first length, then a power P_2 (mW) is measured at the far end, where:

$$P_2 \text{ (mW)} = P_1 \text{ (mW)} \times k = P_0 \text{ (mW)} \times k^2$$

$$\frac{\text{detected power}}{\text{launched power}} = \frac{P_2 \text{ (mW)}}{P_0 \text{ (mW)}} = k^2$$

A further length l would mimic this behaviour such that:

$$P_3 \text{ (mW)} = P_0 \text{ (mW)} \times k$$

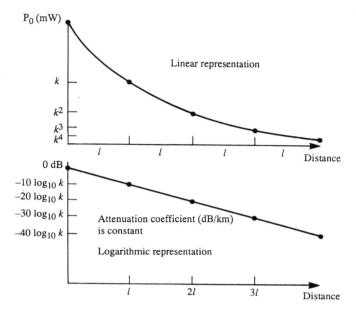

Figure A.2 *Linear versus logarithmic attenuation*

Thus the relationship between detected power and launched power is linear, as shown in Figure A.2. Therefore to calculate the absolute power at any point requires knowledge of the absolute power elsewhere and the ability to utilize the correct scaling factor. To reduce the need for scaling, a logarithmic treatment of power has been adopted such that:

$$\text{Attenuation}_l \text{ (dB)} = -10 \log_{10} (P_1/P_0) = 10 \log k = c$$
$$\text{Attenuation}_{2l} \text{ (dB)} = -10 \log_{10} (P_2/P_0) = 10 \log k^2 = 2c$$
$$\text{Attenuation}_{3l} \text{ (dB)} = -10 \log_{10} (P_3/P_0) = 10 \log k^3 = 3c$$

In this way, a fiber which transmits only 50% of launched power over a kilometre is defined as a 3 dB/km fiber. Two kilometres of such a fiber will lose 6 dB. In this way all fiber losses become additive rather than requiring complex multiplication (see Figure A.3).

Some typical dB (deci-Bel) losses are shown in Table A.1 together with their ratio equivalents.

The general equation below can be applied to all passive network components (joints, connectors, splitters, cables) with the result that system design is much simplified.

$$\text{attenuation (dB)} = -10 \log_{10} \frac{P_{\text{out}}}{P_{\text{in}}}$$

316 Fiber Optic Cabling

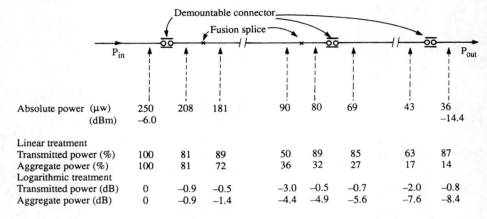

Absolute power (μw)	250	208	181	90	80	69	43	36	
(dBm)	−6.0								−14.4
Linear treatment									
Transmitted power (%)	100	81	89	50	89	85	63	87	
Aggregate power (%)	100	81	72	36	32	27	17	14	
Logarithmic treatment									
Transmitted power (dB)	0	−0.9	−0.5	−3.0	−0.5	−0.7	−2.0	−0.8	
Aggregate power (dB)	0	−0.9	−1.4	−4.4	−4.9	−5.6	−7.6	−8.4	

Total insertion loss = 10 log (36/250) = −8.4 dB

Figure A.3 Comparison of linear versus logarithmic power analysis

Table A.1 Comparison of dB loss and absolute loss/transmission

dB loss	% loss	% transmission
0.1	2.3	97.7
0.2	4.5	95.5
0.3	6.7	93.3
0.4	8.8	91.2
0.5	10.9	89.1
0.6	12.9	87.1
0.7	14.9	85.1
0.8	16.8	83.2
0.9	18.7	81.3
1.0	20.6	79.4
2.0	37	63
3.0	50	50
4.0	61	40
5.0	68	32
6.0	75	25
7.0	80	20
8.0	84	16
9.0	88	12
10.0	90	10

Index

Air blown fiber (ABF), 138, 229
Aspect ratio, 21
Asynchronous transfer mode (ATM), 38, 53
Attenuation, 22, 24, 27, 41

Bandwidth, 22, 31, 45
Bellcore, 124
Biconic optical connector, 115
Blolite®, 138
Blown fiber, 138, 229

Cable cut-off wavelength, 42
Campus cabling, 160
Centralized optical architecture (COA), 190
Chromatic dispersion, 32, 42
Coarse wavelength division multiplexing (CWDM), 162
Core cladding interface (CCI), 20
Critical angle, 17
Cross-phase modulation, 49
Cut-off wavelength, 41

Dark fiber, 158, 173
Decibel, 74
Dense wavelength division multiplexing (DWDM), 162
Differential mode delay, 38
Dispersion compensating fiber, 49
Dispersion shifted fiber, 40
Distributed feedback laser (DFB), 37
Double crucible method, 56

Effective modal bandwidth, 39
EN 50173, 48
Encircled flux requirement, 39
Erbium doped fiber amplifier (EDFA), 49
ESCON®, 163
Ethernet, 31
Extrinsic loss, 24

Fabry Perot laser, 37
Fast ethernet, 31
FC-PC optical connector, 110, 111, 118
Fiber channel, 38
Fiber distributed data interface (FDDI), 31, 115
Four-wave mixing, 49
FOTP, 121
Fresnel reflection, 17
Fusion splicing, 73, 78, 98

Ghosting, 271
Gigabit ethernet, 31
Graded index, 34, 52, 58

Hard clad silica, 48

Ideal fiber model, 92
Index matching gel, 83
Insertion loss, 24, 73
Inside vapour deposition, 59
Interference fit model, 93
Intrinsic loss, 24
Intermodal dispersion, 32
Intramodal dispersion, 32, 42

ISO 11801, 39

Keyed connectors, 103

LC optical connector, 111, 116
Large effective area, 50
Light emitting diode (LED), 37

MPO optical connector, 115
MT-RJ optical connector, 111, 116
MU optical connector, 115
Macrobending, 24, 28
Material absorption, 24
Material scattering, 24
Mechanical splice, 73, 100
MEMS, 312
Microbending, 24, 30
Modal conversion, 36
Modal distribution, 27
Mode field diameter, 42
Mode scrambling, 231
Modified chemical vapour deposition (MCVD), 60
Monomode, 40

Nominal velocity of propagation (NVP), 12
Non zero dispersion shifted fiber (NZDSF), 49
Numerical aperture (NA), 22

Optical budget, 176
Optical code division multiple access (OCDMA), 313
Optical time domain reflectometer (OTDR), 222, 267
Outside vapour deposition (OVD), 60
Overfilled launch, 37

Perflourinated plastic, 52
Pigtail, 170
Photonic bandgap, 51
Plastic clad silica (PCS), 66
Plastic optical fiber (POF), 52, 67
Plasma chemical vapour deposition (PCVD), 60
Plenum, 142

Polarization mode dispersion (PMD), 45
Preform, 56
Primary coated optical fiber (PCOF), 72, 132
Primary coating, 70

Radiation hardness, 68
Rayleigh scattering, 25
Refraction, 15
Refractive index, 12
Restricted mode launch, 37
Return loss, 84
Rod-in-tube, 55

Secondary Coated Optical fiber (SCOF), 133
SG optical connector, 111
Silica, 15
Silicon Dioxide, 25
Singlemode, 39
Sized ferrule, 94
SMA optical connector, 112, 115
Small form factor (SFF), 111
Snell's law, 18
Soliton, 51
Splice joint, 105
ST optical connector, 111, 113, 115
Step Index, 28, 34
Submarine fiber, 51
Subscriber connector (SC), 111, 116

Telcordia®, 124
Telecommunications Industry Association (TIA), 38
Ten gigabit ethernet, 162
Total internal reflection (TIR), 18

Vapour deposition, 59
Vapour axial deposition (VAD), 60
VCSEL, 38

Wavelength division multiplexing (WDM), 49
Windows of operation, 31

Zero dispersion point, 49